FOOD AND AGRICULTURE ORGANIZATION OF THE UNITED NATIONS
Rome, 1999

Editing, design, graphics and desktop publishing:
Editorial Group
FAO Information Division

The designations employed and the presentation of material in this publication do not imply the expression of any opinion whatsoever on the part of the Food and Agriculture Organization of the United Nations concerning the legal status of any country, territory, city or area or of its authorities, or concerning the delimitation of its frontiers or boundaries.

FAO Documentation Group Cataloguing in Publication Data

FAO, Rome (Italy)
State of the World's Forests 1999
ISBN 92-5-104193-8

(State of the World's Forests)
ISSN 1020-5705

1. Forest resources 2. Forestry development
3. Forestry policies 4. Forest protection 5. World
I. Title II. Series

FAO code: 36 AGRIS: K10

Foreword

The *State of the World's Forests* (SOFO) provides information on the current status of the world's forests, as the name implies, but also discusses recent major policy and institutional developments, future directions in forestry and external influences on the sector. The purpose is to make current, reliable and policy-relevant information available to policy-makers, foresters and other natural-resource managers, academics, the forest industry and civil society. In doing so, it is hoped that SOFO will help facilitate informed discussion and decision-making.

SOFO was conceived in 1994 in response to a growing demand for reliable information on the world's forests and the absence of periodic reports giving a global and comprehensive view of forestry. The FAO Forestry Department consequently undertook to publish such a report every two years. The first SOFO was published in 1995. It was followed by a more comprehensive edition in 1997, which attempted to provide an overall view of the forest sector and to put forestry into a longer-term perspective by examining trends from 1970 to the present and by looking ahead to 2010. SOFO 1997 serves as a benchmark document. SOFO 1999 takes a different approach: it concentrates on significant events and developments of the last two years and focuses on a limited number of selected topical issues meriting a more in-depth look.

SOFO 1999 examines developments in the sector from the perspective of sustainable forest management. It sheds light on policy reforms, institutional developments, international conventions, economic changes and other factors affecting forests, forest resources and their provision of social and environmental services. SOFO 1999 addresses several questions: Are the changes taking place in forestry today bringing the world's forests closer to being sustainably managed? Do these changes increase forests' contribution to sustainable development? Are international initiatives to support this process helping to move the sector in the right direction?

Information is a theme that runs through this issue of SOFO. Information and its analysis are the basis for accurate planning and appropriate policy-making. SOFO 1999 takes a closer look at the need for different types of forest information, the scope of databases, national and international capabilities to collect and analyse information and international efforts to make it more accessible. Other topics addressed by SOFO 1999 include changing sources of wood supply, increased people's participation in forest planning and management, and formal and informal instruments for achieving sustainable forest management.

The FAO Forestry Department hopes that SOFO will succeed in increasing awareness of key issues facing the forest sector today and in providing readers with valuable information. We hope that this will contribute to national, regional and international efforts to work towards sustainable forest management, and thus towards sustainable human development.

As with the previous two reports, readers are encouraged to send FAO their comments on SOFO 1999. Additional or updated information that may be incorporated into SOFO 2001 would be most welcomed.

M. Hosny El-Lakany
Assistant Director-General
FAO Forestry Department

Contents

PART III
THE INTERNATIONAL DIALOGUE AND INITIATIVES ON FORESTS

PART IV
FORESTRY IN REGIONAL ECONOMIC GROUPS

Acknowledgements

SOFO 1999 owes its existence to the efforts of many individuals both within and outside FAO. Overall coordination and editing of the publication was carried out by S.M. Braatz. The following FAO staff members and consultants were involved in collecting data, drafting sections or reviewing the document: G. Allard, J. Anderson, J. Ball, J. Balsiger, S. Bass, L.S. Botero, J. Bourke, S.M. Braatz, C. Brown, G. Bull, C.M. Carneiro, F. Castañeda, L. Christy, J. Clément, A. Contreras, R. Davis, P. Durst, C. Eckelmann, M. Hosny El-Lakany, L. Ferroukhi, T. Frisk, T. Hofer, M. Gauthier, S. Hald, R. Heinrich, H. Hilmi, V. Johnston, P. Koné, J. Lahaussois, M. Laverdière, J. Lindsay, L. Lintu, L. Ljungman, P. Lowe, G. Lund, Q. Ma, M. Malagnoux, M. Martin, D. McGuire, A. Mekouar, T. Michaelsen, M. Morell, H. Ortiz-Chour, F. Padovani, C. Palmberg-Lerche, M. Paveri, F. Pontecorvi, C. Prins, H. Qwist-Hoffmann, P. Qwist-Hoffmann, R. Romeo, L. Russo, K.H. Schmincke, E.H. Sène, O. Serrano, D. Shallon, A. Sherwood, P. Sigaud, H. Simons, O. Souvannavong, K. Thelen, M.A. Trossero, P. Vantomme, K. Warner, A. Whiteman, M.L. Wilkie and D. Williamson.

A number of outside collaborators contributed various pieces. H. Gillett (World Conservation Monitoring Centre, United Kingdom) prepared the piece on assessing biological diversity at the species and ecosystem level. B. Schlamadinger and G. Marland (Oak Ridge National Laboratory, United States) and C. Leining and B. Braatz (ICF Incorporated, United States) contributed the piece on forests' role in mitigating global climate change. D. Heuer (Precious Woods Ltd., Switzerland) prepared the box on certification of Precious Woods. D. Barron (Canadian Pulp and Paper Association) and H.S. Leng (Malaysia Timber Council) provided input for the piece on private-sector initiatives. N. Dudley (Equilibrium, United Kingdom) contributed the piece on NGOs for the section on initiatives in the international forestry debate.

The following members of the Internal Advisory Committee provided technical oversight in planning and reviewing the document: J. Anderson, J. Ball, J. Bourke, S. Dembner, M. Paveri, P. Vantomme and K. Warner.

FAO wishes to acknowledge the extremely valuable guidance given by the members of SOFO's External Review Committee: D. Barron, S. Bass, M. Dourojeanni, M. Goumandakoye, H. Gregersen, M. Hadley, J.P. Lanly, J. Maini, O. M'Hirit, C. Prins, M.N. Salleh and M. Simula. D. Kneeland also provided helpful comments.

Many thanks are due to S. Dembner and to the staff of the Editorial Group, FAO Publishing and Media Support Service for production and publication support. L. Ransom and E. Rubini provided essential secretarial assistance.

Executive summary

The *State of the World's Forests* (SOFO) is produced by FAO every two years for the purpose of making current, reliable and policy-relevant information available to policy-makers, foresters and other natural resource managers, academics, forest industry and the public. SOFO 1999, the third in the series, examines the current status of the world's forests, recent policy and institutional developments and external impacts on the sector. It concentrates on significant events and developments of the 1997-1998 period and on selected topical issues.

This two-year period was marked by extremes: record high global temperatures; devastating floods and severe droughts; widespread forest fires and heavy ice storms; and rapid global economic growth abruptly disrupted by the Asian economic crisis. Forests have felt the impact of these climatic and economic phenomena.

Significant changes have also been driven by developments within the forest sector. Policy-makers and forest managers have been responding to changing national priorities and to international commitments made at, and since, the United Nations Conference on Environment and Development (UNCED) in 1992, aimed towards sustainable management of all types of forests. A broader approach to forest management is being sought, balancing social, economic and environmental objectives. The concerns of a wider range of interest groups are being taken into consideration, and local communities are increasingly becoming directly involved in forest planning and management.

SOFO 1999 discusses the state and change of forests worldwide, new developments in forest management and issues and trends related to forest goods and services. It also examines recent policy trends and institutional changes which are having an impact on how, for what purposes and by whom forests are being managed. This global view of forestry today is augmented by information about the forest sector at the regional and subregional levels.

CHANGES IN FOREST COVER AND CONDITION

The latest global figures on forest cover indicate that in 1995 there were 3 454 million hectares of forest (including natural forests and forest plantations) worldwide. Between 1990 and 1995, the total area of forests decreased by 56.3 million hectares – the result of a loss of 65.1 million hectares in developing countries and an increase of 8.8 million hectares in developed countries. Major causes of forest cover change include conversion of forests to agricultural land and large infrastructural development in developing countries, and forest growth on abandoned agricultural land in developed countries.

The many causes of forest degradation include overharvesting of industrial wood and fuelwood, overgrazing, fire, insect pests and diseases, storms and air pollution; of these, forest fires were the most visible in 1997-1998. These were the worst two years for wildfires and forest fires in recent times. Millions of hectares of forest burned. All regions of the world were affected and nearly all types of forests burned. National disasters were declared, and national and international resources were mobilized to fight the fires. Vast areas of forest burned in Brazil, Indonesia, Mexico and the Russian Federation. Although the exact areas are still unknown, estimates are 2 million hectares in Indonesia in 1997 and perhaps more in 1998, and 2 million hectares each in Brazil and the Russian Federation in 1998. While droughts associated with the unusually strong El Niño weather pattern contributed to the increased number, size, intensity and duration of fires, land use practices, mainly in agriculture and forestry, were clearly major causes of many fires. The fires have catalysed national and international initiatives to encourage appropriate policy responses and operational measures for preventing and controlling fires in the future.

TRENDS IN FOREST MANAGEMENT

Forest management objectives, management practices and the managers themselves have changed dramatically in the past decade in many countries. Increased emphasis on environmental values has resulted in significant expansion of the protected area systems (including forest lands) of some countries over the past few years. Greater emphasis on management of natural forests for multiple purposes and increased attention to environmental factors have led, in some places, to reduced intensities of timber harvesting and changes in management practices. Logging bans in natural forests have been announced in several countries. The increased emphasis on the use of natural forests for environmental protection, conservation of biological diversity and recreation has increased the reliance on forest plantations and other sources of wood supply in some countries.

Efforts to improve the stewardship of forests designated for wood production include silvicultural improvements and the adoption of environmentally sound timber harvesting practices. Operational guidelines and codes of practice for forest management have been adopted in Europe and North America. New silvicultural systems have been developed for temperate and boreal forests in North America and for tropical forests. Codes of practice and guidelines for forest harvesting have recently been drafted for tropical forests, most notably in the Asia and the Pacific region.

Changing patterns of forest ownership, increasing support for locally based forest management and greater participation of the private sector in forest management have influenced how forests are managed and by whom. Changes leading to increased local forest management in various countries include:

- a proliferation of new mechanisms for the devolution of forest management to local communities or user groups;
- increased recognition of the historical territorial claims of local peoples;
- restoration of the lands (including forests) of dispossessed communities and individuals.

SOCIAL AND ENVIRONMENTAL SERVICES OF FORESTS

Recognition of the social and environmental services of forests – such as mitigation of global climate change, conservation of soil and water resources, enhancement of agricultural systems, conservation of biological diversity, improvement of urban and peri-urban living conditions, protection of natural and cultural heritage, generation of employment and provision of recreational opportunities – continues to grow.

As the emphasis on these important services of trees and forests has increased, more attention has been focused on forests in fragile ecosystems – including mountains, drylands and small island States – and in countries with low forest cover. Although these forests generally have relatively low timber value, their important social and environmental functions have been increasingly recognized in management decisions. Forests' role in water conservation is expected to be given more prominence as international attention increasingly turns to freshwater resource issues.

The role of forests in mitigating global climate change was given greater prominence with the adoption of the Kyoto Protocol of the Framework Convention on Climate Change in 1997 and the deliberations of the fourth Conference of the Parties in Buenos Aires, Argentina in November 1998. The Kyoto Protocol establishes legally binding commitments for reduced emissions of greenhouse gases from industrialized countries and permits the use of a limited list of activities in the land use change and forest sector to meet these commitments. The protocol thus provides industrialized countries incentives to invest, in both their own and other countries, in forestry activities which increase carbon sequestration and/or reduce carbon emissions. While many details of the Kyoto Protocol remain to be clarified, if ratified it should offer possibilities for investment in forestry.

FOREST PRODUCTS

Meeting the needs for wood and non-wood forest products while at the same time fulfilling demands for environmental and social services from forests remains a major challenge.

While wood is the predominant commercial product from most forests, increased attention is being paid to non-wood forest products (NWFPs), whose main importance currently lies in their contribution to household and local economies, particularly among the poor in developing countries. Recent regional and international fora have addressed the needs to conserve non-wood resources while ensuring local peoples' equitable access to and use of the resources, to improve market information and to develop appropriate and fair pricing mechanisms for NWFPs (including royalties on intellectual property rights). SOFO 1999 focuses specifically on medicinal plants, which as a group are among the most valuable NWFPs from forests. Most medicinal plants gathered from the wild come from forest lands.

Fuelwood and charcoal are expected to continue to be important sources of household energy in developing countries. Moreover, recent policy changes and experiences with wood-based bio-energy programmes in several countries indicate that woodfuels may become increasingly attractive to some countries as industrial energy sources.

Developed regions continue to dominate heavily in production and consumption of industrial wood products. FAO has projected that demand for industrial roundwood will increase by 1.7 percent annually between now and the year 2010, driven both by population increases and economic growth. While recent studies suggest that supplies are sufficient to meet this demand, the situation will vary among countries and will depend greatly on market conditions, government policies, technological improvements and human resource development. Production of industrial roundwood is expected to continue to exceed consumption in all regions except Asia, which will continue to rely on imports to make up the difference.

Recent trends which are expected to continue to help in meeting the increasing demand for processed wood products include:

- improved management of forests designated for wood production;
- increased production from plantations and from trees outside forests, particularly on agricultural land;
- new forest product technologies and gains in efficiency in wood processing;
- greater use of recovered paper and wood processing residues;
- increased use of wood and fibre from "non-forest" species (e.g. rubber, oil palm) in forest industries.

Trade will help balance deficits of wood fibre in one place with surpluses elsewhere.

The Asian economic crisis, which began in mid-1997, has seriously disrupted forest products trade in the Asia and the Pacific region. The main impacts on countries within the region have been:

- reduced demand for all forest products;
- increased competitiveness of exporting countries whose currencies have depreciated;
- falling prices for forest products throughout the region;
- reduced earnings in the forest sector, resulting in reduced harvests, mill closures and workforce lay-offs.

While Asian wood markets have been the most severely affected, the effects have spread outside the region to countries that either depend heavily on, or compete with, Asian markets. Wood exporters from Africa to New Zealand and from North to South America have been affected.

Certification remains a high-profile but complex and often controversial issue. Various international, regional and national certification systems have been developed. The area of forests certified has increased considerably in the past two years, although significant volumes of certified products apparently are not yet entering the market. It is unclear whether the demand for certified wood will increase and whether certification will, in fact, significantly contribute to improved forest management where deforestation is greatest, i.e. in developing countries.

RECENT CHANGES IN FOREST POLICY, LEGISLATION AND INSTITUTIONAL FRAMEWORK

Evident trends in forest policy reform include attempts to privatize State forest resources and public forest-based companies, to decentralize certain functions of central government administrations and to eliminate some "perverse subsidies", including the underpricing of forest concessions.

Revision of forest-related laws around the world has accelerated significantly in recent years. The legal changes reflect commitments to:

- promote local forest management;
- enhance the environmental functions of forests;

- reinforce forest management planning, increasing consideration of environmental and social issues in management plans and supporting public participation in the planning process;
- increase transparency in awarding forest concessions.

Institutional changes in the forest sector – particularly privatization and the decentralization and devolution of forest management responsibilities to the local level – have closely paralleled these policy and legislative trends. Decentralization and privatization are also affecting forest research systems. Forest education and extension are responding to changing objectives for forest management, new forest owners and/or managers and the forest-related concerns of an increased number of interest groups.

INTENSIFIED EFFORTS TO DEVELOP FOREST-RELATED INFORMATION

Many organizations have made substantial efforts over the past few years to collect and disseminate global forest-related information. Major initiatives have been undertaken in forest cover assessment and mapping, forest resources assessment, study of wood supply and data collection on biological diversity. Complementing these global efforts are various initiatives to collect and analyse information at the regional level and to strengthen national capacity in the collection, analysis and use of forest-related information.

INTERNATIONAL INITIATIVES ON FORESTS

Many recent international initiatives are aimed at fostering sustainable forest management. More than 150 countries are currently participating in international processes to develop and implement national-level criteria and indicators for sustainable forest management. Various efforts at the forest management unit level are being coordinated by the Center for International Forestry Research (CIFOR), the International Tropical Timber Organization (ITTO) and the African Timber Organization (ATO). Other international initiatives include the establishment and management of model and demonstration forests, ITTO's "Year 2000 Objective" and the G8 countries' action programme on forests which was approved in May 1998. Developments in the implementation of the three international conventions arising from UNCED (the Framework Convention on Climate Change, the Convention on Biological Diversity and the Convention to Combat Desertification) and the International Tropical Timber Agreement provide additional support for certain elements of forest management. Non-governmental organizations and the private sector continue to be active in the international forest debate and in efforts to improve forest management at the field level.

The post-UNCED international dialogue on forests led by the Intergovernmental Panel on Forests (IPF) from 1995 to 1997 resulted in over 100 negotiated proposals for action. Work continues under the Intergovernmental Forum on Forests (IFF), established in mid-1997. In 2000, IFF will present its final report to the United Nations Commission on Sustainable Development on its three major areas of work:

- the implementation of IPF's proposals for action;
- issues on which international consensus has yet to be achieved;
- the identification of possible elements of, and work towards, consensus on international arrangements and mechanisms for the management, conservation and sustainable development of forests.

The issue of international instruments for support of forests first arose in 1990. Since then, a number of such instruments, both legally and non-legally binding, have been developed. Consensus on whether additional international instruments are needed, and if so, what they would cover, remains elusive. Discussion of whether to rely on existing mechanisms, to enhance them or to develop additional ones – such as a legally binding instrument on all types of forests – promises to remain high on the international forest agenda for at least the next few years. ◆

PART I
SITUATION AND PROSPECTS FOR FOREST CONSERVATION AND DEVELOPMENT

The status of forest resources

Information on the status of the world's forests – their extent, location, type and condition – is important for efforts to improve forest management worldwide and for assessments of forests' ability to provide the goods and services demanded of them. The following discussion focuses on two subjects: the area and condition of forest resources worldwide, and the status of the information base on global forest resources, particularly concentrating on information on forest area and forest-based biological diversity.

STATUS OF FOREST RESOURCES AND RECENT DEVELOPMENTS IN FOREST CONDITION

Forest cover[1]

The area of the world's forests, including natural forests and forest plantations, was estimated to be 3 454 million hectares in 1995, or about one-fourth of the land area of the Earth. About 55 percent of the world's forests are located in developing countries, with the remaining 45 percent in developed countries (Figure 1) (see Table 2 of Annex 3 for country data). The world's forests are almost equally divided between tropical/ subtropical forests and temperate/boreal forests. Only about 3 percent of the world's forests are forest plantations. The remaining 97 percent are natural or semi-natural forests.

Change in forest cover

Data published in the *State of the World's Forests 1997* (FAO, 1997d) provide a picture of the trends in forest cover over a 15-year interval (1980-1995) and allow a comparison between the 1980-1990 and 1990-1995 periods. Between 1980 and 1995, the extent of the world's forests (including both natural forests and forest plantations) decreased by some 180 million hectares. There was a net increase of 20 million hectares in developed countries, but a net loss of 200 million hectares in developing nations. The change in forest area by region between 1980 and 1995 is shown in Figure 2.

Between 1990 and 1995, there was an estimated net loss of 56.3 million hectares of forests worldwide. This represented a decrease of 65.1 million hectares in developing countries and an increase of 8.8 million hectares in developed countries. Although the global loss of forests was still very high, the figures suggest that the rate of deforestation might be slowing. The estimate of forest cover change in natural forests of developing countries (which is where most deforestation is taking place) was an annual loss of 13.7 million hectares between 1990 and 1995, compared with 15.5 million hectares per year over the decade 1980-1990. It will be difficult to know if this is a trend, however, until a comparable global data set is available from the Global Forest Resources Assessment 2000 (see following discussion).

The major causes of change in forest cover in the tropics appear to be expansion of subsistence agriculture in Africa and Asia and large economic development programmes involving resettlement, agriculture and infrastructure in Latin America and Asia (FAO, 1996a). The net increase in forest area in developed countries is largely a result of afforestation and reforestation, including natural regrowth on land abandoned by agriculture. This increase has more than compensated for the clearing of some areas of forest in various developed countries, mainly for urban expansion and infrastructure development.

Forest condition: fire, ice and forest pests

The causes of forest degradation vary from place to place, and the magnitude and duration of the

[1] The *State of the World's Forests 1997* provided the most recent data available on global forest cover. As no new global data set on forest resources will be available until the results of the Forest Resources Assessment 2000 study are published, a summary of the 1995 figures published in SOFO 1997 is provided in this section.

FIGURE 1
Forest areas by main regions in 1995

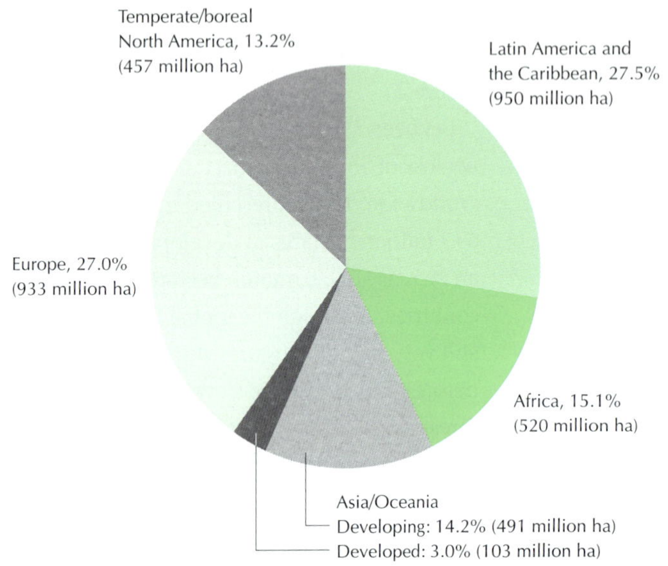

Total area (1995): 3 454 million ha
Developed countries: 1 493 million ha
Developing countries: 1 961 million ha

FIGURE 2
Forest area in 1995 as compared with 1980

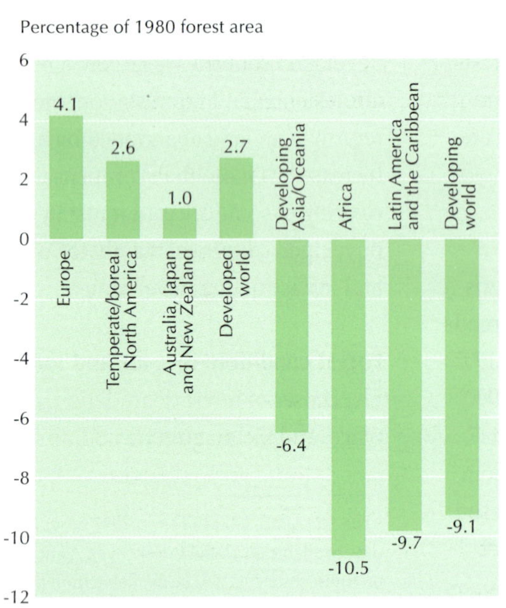

Note: Data exclude the countries of the former Soviet Union.

effects are often difficult to assess. The causes include insect pests and diseases, fire, overharvesting of industrial wood and fuelwood, poor harvesting practices, overgrazing, air pollution and extreme climatic events, such as storms. Particularly noteworthy causes of forest degradation in the 1997-1998 period were the dramatic wildfires that affected forests throughout the world, the 1998 ice storms in the United States and Canada and new insect pest and disease outbreaks.

Globally, 1997 and 1998 were the worst years for wildfires and forest fires in recent times. Although forest fires occur every year in the arid and semi-arid zones of the world, nearly all types of forests burned in 1997-1998, even some tropical rain forests which had not burned in recent memory. Droughts associated with the El Niño weather pattern (see Box 1) turned moist forests into drier habitats and increased the flammability of forest vegetation, thus increasing the number, frequency, size, intensity and duration of fires.

In 1997, wildfires raged in Indonesia, Papua New Guinea, Australia, Mongolia, the Russian Federation, Colombia, Peru, Kenya, Rwanda and other parts of Africa. By mid-1998, fires were reported in Indonesia, the Amazon, Mexico and Central America, the United States, western Canada, far-eastern Russia and parts of Europe. National disasters were declared in many of these places and national and international resources were mobilized to fight the fires. Neighbouring countries and international and non-governmental organizations (NGOs) all responded. In March 1998, the Secretary-General of the United Nations requested the United Nations Environment Programme (UNEP) to coordinate the UN system's response to the situation arising from the outbreak of forest fires in Indonesia.

Few global figures are available on the extent of the 1997-1998 fires and associated loss of life, economic damage and environmental impacts.

The largest areas of forests burned in Brazil and Indonesia. Low rainfall in much of the Amazon attributed to the El Niño weather pattern contributed to a prolonged fire season (beyond the usual July to early October period) and an unusually high number of fires. In 1997, over

BOX 1

The 1997-1998 El Niño southern oscillation: El Niño and La Niña

The El Niño Southern Oscillation (ENSO) is a periodically occurring oceanic-atmospheric phenomenon. It consists of El Niño, a "warm phase" or a large-scale warming in the equatorial Pacific Ocean, and La Niña, a "cool phase" in which surface waters of the central Pacific Ocean are colder than normal. These phases are closely associated with changes in atmospheric conditions (Southern Oscillation). ENSO events occur on average every four years and typically last 14 to 22 months. They exhibit certain characteristic rainfall and temperature anomalies. The El Niño that began in March 1997 and continued until mid-1998 is believed to have been one of the most severe in recorded history. It had a pronounced impact on weather and climate around the globe. While some parts of the world received more rain than normal, abnormally dry conditions occurred across northern Australia, Indonesia and the Philippines in both seasons and in southeastern Africa and northern Brazil during the northern winter season, and Indian monsoon rainfall was lower in the northern summer season. The record-high global temperatures in 1998 were also believed to be linked to El Niño. La Niña started to develop in mid-1998. Climatic anomalies in 1998 which may be associated with La Niña include dry conditions in parts of South America; wetter than normal conditions in northern Australia and the Philippines; above-normal rainfall during the southwest monsoon in India; increased hurricane activity in the Caribbean and Central America; dry spells in parts of Argentina and Chile; above-normal rainfall in southern Africa, with the exception of Zimbabwe; and possibly drier than normal conditions in the Horn of Africa.

2 million hectares of rain forest in Brazil burned (Schemo, 1998). Analysis of satellite data from the United States National Oceanic and Atmospheric Administration showed an increase of over 50 percent in the number of fires from July to November 1997 compared with the same period in 1996, and an 86 percent increase in a 100-day period (June to early September) in 1998 compared with the same period in 1997 (Schwartzman, 1998). Most of the fires occurred in Mato Grosso and Pará states.

Some Brazilian forests of particular ecological or cultural significance were affected. In March 1998, fires burned over 600 000 ha of rain forest in Roraima, including parts of the Yanomami Indian reserve, near the border with Venezuela (Schemo, 1998). In late September 1998, raging fires destroyed a large area of Brasilia National Park – a sanctuary for rare species from Brazil's central savannah region – killing wildlife and smothering the Brazilian capital with smoke. Earlier in the month, fire in the state of Mato Grosso threatened to move into Xingu National Park, home to 17 indigenous groups, until rains extinguished the fire.

In Indonesia the fires of 1997-1998 burned millions of hectares in Sumatra and Kalimantan. The exact area is still unknown. One estimate is that about 2 million hectares (including savannah with grassland) burned in 1997 alone. Several organizations have begun the lengthy and complex task of interpreting satellite images to determine the total area burned (Schweithelm, 1998). Large quantities of smoke generated by ground fires fed by slow burning fuels affected neighbouring countries, influencing human health, interfering with transportation systems and disrupting the multi-million-dollar tourist industry, all of which contributed significantly to the economic and social cost of the fires. Many underground fires continued to burn into mid-1998 in natural peat and coal beds, threatening new outbreaks of fire.

A representative sample of forest fires elsewhere in the world includes the following:

- Fires in Mexico and Central America burned a reported 1.5 million hectares. These generated large quantities of smoke which blanketed the region and spread into the United States as far as Chicago.
- From January to June 1998, about 13 000 fires burned in Mexico alone, consuming nearly

500 000 ha and killing more than 70 fire-fighters and local residents.

- Between December 1997 and April 1998, more than 13 000 fires burned in Nicaragua, the most in any Central American country, destroying vegetation on more than 800 000 ha of land. The Nicaraguan Ministry of Environment and Natural Resources (MARENA) recorded over 11 000 fires in the month of April 1998 alone.
- In the southeastern United States, severe fires in Florida in 1998 burned a reported 200 000 ha of forest by May (*The Economist*, 4-10 July 1998).
- More than 150 000 ha of coniferous forest and farmland were burned in various parts of Greece in August 1998, including the black pine forest on Mount Taygetos, site of more than 160 endemic species and 36 endangered species of fauna (*Guardian Weekly*, 30 July 1998).
- In July 1998, devastating forest fires affected more than 100 000 ha in far-eastern Russia. Coniferous forest burned in more than 150 places around Vladivostok, Sakhalin and Kamchatka Peninsula. Fires in the southwest of the Volgograd region destroyed 9 000 ha of forest, at an estimated cost of US$6 million (BBC World Service, 27 July 1998). In September, fires swept across Russia's Pacific island of Sakhalin, burning over 25 000 ha by the end of the month (Reuters News Service, 21 September 1998).

Active media coverage of the unusually large fires in Indonesia, the Amazon and Mexico increased public awareness of these predominantly man-made environmental disasters. The smoke from the fires, which endangered public health and economies beyond national borders, also helped focus public attention on the fires and the need to deal with policy issues related to fire outbreaks.

The use of fire is an integral part of land management in both agriculture and forestry. Some forests are ecologically adapted to fire, and prescribed burning is an important component of forest management practices in some places. However, there is ample evidence that a large proportion of the wildfires that occur each year are caused by human intervention. A seven-year field study of fire in the Amazon carried out by the Woods Hole Research Center, United States, shows that about half of the wildfires in the region were caused by fires set by farmers and ranchers to clear old cattle pasture and to burn newly cut forest which subsequently burned out of control. The effects of drought in 1997-1998 were aggravated by shifting cultivation (particularly slash-and-burn agricultural practices), pasture management, improperly executed timber harvesting operations and large-scale land clearance under agricultural conversion schemes. A considerable number of fires that burned around the world in 1997-1998 (as in other times) were cases of arson. The fires of 1997-1998 clearly highlight the need to improve planning and management of agricultural land use practices to reduce the risk of wildfires; to address deficiencies in forest management systems; and to examine policies and regulations related both to conversion of forest land to agriculture and to improvement of forest stewardship.

The fires of 1997 and 1998 have stimulated various international efforts related to fires. A global system of early warning to indicate the potential fire risk related to climatic conditions is being investigated by several international organizations, including the World Health Organization (WHO), the World Conservation Union (IUCN), UNEP and FAO. A national system for advance warning by radio linked to meteorological forecasts from satellite imagery has been successfully tried in Burkina Faso. WHO has produced guidelines for forest fire emergencies. Many national and international meetings were organized in 1998 for fire-fighting and health experts, potential donors and, perhaps more significant, policy-makers to address the control, effects and underlying causes of fires. UN agencies that organized fire meetings in 1998 include UNEP and the United Nations Office for the Coordination of Humanitarian Affairs (Geneva, Switzerland, April 1998), WHO and the Pan-American Health Organization (Lima, Peru, August 1998) and FAO (Rome, October 1998, on public policies affecting forest fires).

Apart from the fires, 1998 was notable for another type of disaster which severely affected forests: ice. A series of ice storms in January 1998 resulted in extensive damage to forests in the northeastern United States and eastern Canada (see Box 2). An estimated 10 million hectares of forests and trees were affected, extending from Ontario to Quebec and New Brunswick in Canada and across northern New England in the United States (Irland, 1998).

While less visible than the wildfires and ice storms within the last two years, recent outbreaks of insect pests and diseases in forests may result in substantial economic loss and environmental damage. Although insect pests and diseases are often integral components of forest dynamics, severe outbreaks can be damaging and require costly control efforts. Their incidence and economic impact may be highest on plantations established for the purpose of wood production. Some new pest outbreaks reported in the last few years are noted in Box 3.

Climatic extremes associated with El Niño in 1997-1998 (see Box 1) are thought to be linked to some of the current severe pest outbreaks affecting forests and trees. For example, a serious outbreak of gypsy moth (*Lymantria dispar*) which is posing a risk to northern European forests is believed to have been exacerbated by increased temperatures and prolonged drought conditions over the last two years. The outbreak, which started in 1996 and was still ongoing in mid-1998, is affecting thousands of hectares of valuable hardwoods (particularly oaks) in Bulgaria, Croatia, Romania and Serbia.

STATUS OF EFFORTS TO ASSESS GLOBAL FOREST RESOURCES

Forest resources assessment

Forest resources assessments generally provide information on forest extent and location, types, condition, wood volume, and / or biomass. They may be done on a subnational, national, regional or global level. Global assessments are done in one of two ways: using a bottom-up approach which harmonizes and sums national statistics, or using a top-down approach in which data are generated at the regional, continental or global level. National forest inventories are fundamental to bottom-up approaches to forest resources assessment (see Box 4).

The major global forest cover assessment efforts are being carried out by FAO, the International Geosphere-Biosphere Programme (IGBP), the European Commission's Joint Research Centre (JRC) and the World Conservation Monitoring Centre (WCMC). FAO and WCMC mainly use a bottom-up approach, which provides the most

BOX 2

Effects of the ice storm in the northeastern United States in January 1998

In January 1998, in what meteorologists called a "100-year event", a series of ice storms blanketed northern New England and New York with up to 7.5 cm of ice. Nearly 7 million hectares of rural forests and urban trees across the states of Maine, New Hampshire, Vermont and New York were affected. Some 2 million hectares were severely damaged. Hardwood species suffered most. Estimates for natural resource losses exceeded US$1 000 million. The effects of the storm remain: thousands of kilometres of forest roads and trails are littered with debris; many rural landowners will lose forest income; management plans may be obsolete and need costly revision; and debris increases the risk of insect pest outbreaks and fire.

Source: USDA Forest Service, 1998.

BOX 3

Recent insect pest and disease outbreaks

In the Philippines, new insect pest outbreaks include the moths *Nyctalemon* spp. on *Endospermum peltatum* and *Xyleutes* spp. on *Gmelina arborea,* which are affecting plantations, and the scolytid beetle *Dryocoetiops laevis* on *Dipterocarpus grandiflorus* in secondary forests. The giant wood moth (*Endoxyla* spp.) and the cerambycid beetle (*Phoracantha* spp.) are becoming significant new threats to eucalyptus plantations in Queensland, Australia. The Asian long-horned beetle, *Anoplophora glabripennis,* is affecting urban hardwoods in New York and Chicago, United States. The gypsy moth (*Lymantria dispar*) is affecting thousands of hectares in eastern Europe and Turkey. New disease problems include *Phythopthora* spp. on *Elmerrillia* seedlings in Indonesia and root rot of *Acacia mangium* in the Philippines caused by *Phellinus noxius.*

BOX 4
Importance of forest inventories as a basis for sustainable forest management

Forest inventories and statistical systems serve as the foundation of sound policies to support sustainable forest management. Integrating social, economic and environmental concerns into forest-sector planning requires a great deal of information on forests, across the landscape and through time. Studies repeatedly come to the same conclusion: much better information is needed on the quantity, quality and use of forests. Despite growing demands for better information, investments in even basic forest inventories are declining. Many countries have not conducted a comprehensive and statistically sound forest inventory since the 1970s or early 1980s. In other countries, physical inventories are declining in frequency and intensity and are being replaced by modelling. New methods, such as remote sensing, are expanding the ability to observe large changes in land cover. Still, without reasonably current forest inventories, it is increasingly difficult to assess change in forest quality and function and to draw useful conclusions about the sustainability of use.

detail at the country level. IGBP and JRC use a top-down method, which provides spatial information on forest cover over broad areas. The FAO, JRC and IGBP assessments are ongoing, long-term programmes.

Currently available information. FAO, in collaboration with other partners, has been conducting global forest resources assessments about every ten years since 1947. The latest Forest Resources Assessment (FRA), which was done for 1990, was carried out jointly by FAO, the United Nations Economic Commission for Europe (ECE) and their member countries (FAO, 1995).[2] FRA 1990 was divided into two parts: one for the industrialized countries (the temperate and boreal forests) and the other for developing countries (mainly tropical and subtropical, but also some temperate forests). ECE and FAO conducted the temperate/boreal forest resources assessment (TBFRA) through the use of a questionnaire. The tropical assessment was based on existing reliable information and a remote-sensing-based sampling design for study of land cover changes. The assessment for developing countries produced information primarily about the area of forest in 1990 and the change since 1980, with data from different years adjusted to these reference points. This information is available by country and by ecofloristic zone. TBFRA includes data on state and change of forest area (which have not been adjusted to a given reference year) and gives additional information on ownership and management status, growing stock, annual growth and fellings. FRA 1990 is currently the most comprehensive source of global-, regional- and national-level information on forest state and change available. It does not, however, show the spatial distribution of forests.

Since FRA 1990, new information requirements have emerged, including location-specific or spatial data which support efforts in sustainable forest management and assessment of forests' environmental functions (e.g. conservation of biological diversity, mitigation of global climate change). As a consequence, several groups, including IGBP, JRC and WCMC, have embarked upon additional global assessments of forest cover and deforestation. All are assessing various aspects of the spatial distribution of forest cover, although none are as comprehensive as the FRA programme or present the same information.

IGBP, an international consortium of scientific organizations, developed a global low-resolution satellite image-mapping project. The work is based on interpretation and classification of advanced very-high-resolution radiometer (AVHRR) satellite imagery which has a 1-km resolution. The IGBP land cover data set and maps and a related Global Land Cover Characterization database are being validated. While the main objective of these projects is not to provide information on forest area *per se*, estimates of forest area may be generated from the database and biomass estimates may be derived from modelling.

The JRC TREES (Tropical Resources and Environment monitoring by Satellites) project,

[2] A partial, interim assessment using modelling was done for forest cover and forest cover change for the reference year 1995; the results were published in SOFO 1997.

established in 1991, develops forest cover assessment techniques for the tropical belt. The project includes:

- a baseline inventory of forest area (using a multispectral 1-km AVHRR data set);
- continuous monitoring which would direct attention to the areas undergoing the most active land cover change (i.e. deforestation);
- analysis of the derived information in terms of spatio-temporal patterns of deforestation through the comprehensive Tropical Forest Information System (TFIS).

A multiyear set of AVHRR data was acquired over the whole tropical belt during the period 1991-1994. This global assessment has been further calibrated using high-resolution image maps for selected sites. The main outputs of this effort include a global tropical forest cover base map at 1-km resolution (the first such forest map at this resolution) and related figures at the regional to national levels.

The second phase (1996-1999) of the TREES project is directed at developing a prototype that can regularly produce relevant and accurate information on the state of tropical forest ecosystems. A panel of international experts has identified "hot spots" where deforestation is the most active. JRC is developing more automatic methods of forest monitoring using a range of remote-sensing and other spatial data and has developed a new sampling scheme to measure changes using different types of high-resolution satellite imagery.

TFIS organizes the data (remote-sensing or cartographic) and current knowledge (reports, field observations, etc.) in a systematic manner. Work is being done to allow an interface between TFIS and information systems of the European Commission and other regional and international organizations.

In 1996, WCMC, in collaboration with the World Wide Fund for Nature (WWF), compiled a World Forest Map showing forest extent and protected areas with forested land. The map is the product of a global geographic information system (GIS) containing more detailed maps and digital files from the best available national and international sources, compiled mainly between the early 1980s and early 1990s by WCMC in collaboration with IUCN. WCMC used a nominal scale of 1:1 million, adjusting input data of varying scales (ranging from 1:100 000 to 1:5 million).

The digital information was used to estimate the area of each major forest type in the world under protection and to provide a baseline for future monitoring of forests. The spatial data have formed the basis for a major statistical analysis of forest protection in the world, carried out by WCMC in conjunction with the Center for International Forestry Research (CIFOR), and have been published on CD-ROM. The digital map is maintained at WCMC.

WCMC recently produced a new set of national-level forest statistics derived from the land cover maps overlaid with the political boundaries of countries. This could provide area estimates at the national level.

WCMC has also developed a GIS database showing an estimate of forest cover some 8 000 years ago for the World Resources Institute (WRI) and WWF. Using the data from WCMC, WRI, under its Frontier Forests Initiative,[3] developed another global map comparing the large tracts of forest lands remaining today with those existing 8 000 years ago. The Frontier Forests Initiative assessed three attributes: historic forest lost, forest condition (the amount of the world's current forest cover remaining in large tracts of intact natural forest) and threats to frontier areas. The WRI/WCMC map may be considered a global assessment of natural forest change over a large time span.

As a follow-up, WRI has launched the major Global Forest Watch project to monitor forest development globally. Currently WRI is working with some 20 NGO partners in Cameroon, Gabon, Canada and Indonesia to map concessions, roads and other developments. The goal is to carry out a similar effort in all major forested countries within the next five years.

Each of the above-mentioned efforts uses different definitions, sources and methods for classifying the vegetation. Their estimates of forest

[3] The Frontier Forests Initiative is a five-year, multidisciplinary effort to promote stewardship in and around the world's last major frontier forests by investment, policy and public opinion.

cover are thus not directly comparable. The quality of information from any assessment depends largely on the quality of the basic input data. The maps from WCMC and WRI were compiled from existing data sources, which could reflect land cover, land use and/or land potential. Consequently, they suffer from the same problems of harmonization as do most bottom-up approaches.

The Global Observation of Forest Cover (GOFC), a new effort of an international partnership of institutions including FAO, will build on previous global forest assessments using earth satellites. GOFC has a goal of stimulating the production of globally consistent products derived from satellite data, supplemented by *in situ* data, by 2002. This will be achieved by facilitating the provision of forestry information and satellite data to organizations such as FAO and UNEP, the Global Climate Observing System (GCOS) and the Global Terrestrial Observing System (GTOS), NGOs and national forest ministries through partnership arrangements with space agencies.

Information to be available through FRA 2000. The FAO Global Forest Resources Assessment 2000 (FRA 2000) will be the most comprehensive and complete assessment yet. As envisioned, FRA 2000 will consist of:

- a common set of definitions and standards for both tropical and temperate/boreal forests;
- a common set of parameters to be assessed, including not only the traditional timber information, but also information, where available, on non-wood forest products, naturalness, biological diversity, protected areas, carbon sequestration functions, forest condition and socio-economic functions;
- a remote-sensing-based "wall-to-wall" global forest map keyed to forest classes used in the assessment, an ecoregion map and a database at 1-km resolution;
- a global sample of satellite data to measure changes in forest cover over the past 20 years.

As the data will be spatially registered, researchers may be able to measure changes in land cover over time; thus a base may be established for improved comparisons for beyond 2010.

FRA 2000 will make further improvements on the data-gathering techniques used in 1990 and will also make use of the data and technologies of other organizations. FAO will use some of the imagery and GIS data sets of IGBP, JRC, the Earth Resources Observation Systems (EROS) Data Center of the United States Geological Survey and others to develop the global forest cover map and database for FRA 2000.

EROS Data Center, FAO and WCMC are developing the Global Forest Map, which will be the first globally consistent small-scale map showing forest cover distribution. The map is being generated through refinement and modelling of IGBP's Global Land Cover Characterization database using advanced image processing and GIS techniques, combined with validation by national experts and existing reliable information.

In the past, an update on global forest resources was possible only every ten years. Institutional development and communication technologies may eventually make it possible to have much more frequent updates and much better quality of data. While technological gains have improved the precision of the estimates and the range of information that can be collected, many countries still need assistance in the planning, execution and financing of their own national forest inventories.

These assessment efforts will continue to provide vital information on the location and extent of forests at the global level. Even more valuable will be the information on net change in extent and condition of forests over time, which addresses the heart of the problem – what is happening to the world's forests and consequently to forests' ability to provide the range of goods and services demanded of them.

Assessments of forest-based biological diversity

Natural forests are arguably the single most important repository of terrestrial biological diversity (diversity of ecosystems, species and genetic resources). Diversity is an essential factor in maintaining forest function. Conservation and management of biological diversity are, therefore, important issues in forest planning. Global, regional and national planning, priority-setting and decision-making related to the conservation of biological diversity all depend on an understanding

TABLE 1
Data availability for the major taxonomic groups

Taxonomic group	Total species	Number of forest-occurring species	Total endemic and/or restricted-range species	Number of forest-occurring endemic species	Total threatened species	Number of threatened forest species
Birds	xxxx[a]	+++	xxxx[a]	+++	xxxx[b]	+++
Mammals	xxxx[a]	+++	xxxx[a]	+++	xxxx[b]	+++
Reptiles	xx[a]	+	xxx	+	xx	+
Amphibians	xx[a]	+	xxxx	+	xx	+
Invertebrates[c]	x	+	x[d]	+	x	x
Trees	xx	+++	x	+++	xx[e]	+++
Palms	xxx[f]	+++	xxx	+++	xx[f]	+++
Ferns	xxx[f]	+++	xxx	+++	xx[f]	+++
Other plants	xx[f]	+	x	+	xx[f]	+

Notes: xxxx data available for all countries; xxx data almost complete; xx data incomplete; x data very incomplete; +++ division into forest/non-forest species needed, being done for FRA 2000; + division into forest/non-forest species needed.
[a] WCMC, 1994.
[b] WCMC, 1996b.
[c] The large number of invertebrates makes them unsuitable for attempting numerical analyses.
[d] Data available for a small number of groups (dragonflies, swallowtail butterflies).
[e] WCMC, 1998a.
[f] WCMC, 1998b.

of the distribution of species and ecosystems, their protection status and the threats to them, and on information on the status of the conservation and use of forest genetic resources.

Species diversity. Knowledge of the world's diversity of species, their distribution and status is very incomplete. Some groups of organisms (e.g. birds) and some parts of the world have been studied more than others. Furthermore, total species diversity is better understood than the diversity of species associated only with forests (see Table 1), as the latter requires an assessment of habitat occurrence or dependence. Evaluation of forest species diversity is meaningful only for the better-known taxonomic groups. Efforts to classify birds, mammals and some plant groups according to forest occurrence are part of FRA 2000.

A detailed understanding of the distribution of species richness within forests requires mapping of species distributions or richness contours. This work has been undertaken several times on a national scale. On a regional scale, progress has been made in mapping total species richness in many groups (e.g. the Mapping African Biodiversity Patterns project of the Danish Centre for Tropical Biodiversity), but as yet these efforts do not incorporate separate analysis of forest species.

Other approaches for evaluating forest species diversity and/or identifying priority areas for its conservation depend on the overlay of different estimates of "biodiversity hot spots"[4] with forest cover data. Such hot spot data sets have been compiled, for example, for endemic bird areas (EBAs), i.e. areas containing at least two restricted-range bird species (those whose distribution covers less than 50 000 km^2) (Bibby *et al.*, 1992). These have been quite accurately mapped, and in the absence of similar data for other taxonomic groups, the EBA data provide a useful measure of the importance of biological diversity for particular geographic areas.

[4] Ecosystems which have a high level of biological diversity and are under threat of destruction.

Other hot spot data sets include WCMC and IUCN's *Centres of plant diversity* (WWF/IUCN, 1994), WWF-US's *Global 200 ecoregions*[5] and Conservation International's *Hotspots*.[6]

Mapping could also be used to identify areas where threatened and endangered forest species are concentrated. This would require categorization of species according to their forest occurrence and, ideally, relatively detailed data on their distributions. Nonetheless, some advances could be gained simply from addressing presence/absence data at the national or provincial level.

A country may have high forest species diversity either because it has a wide range of different forest types each with its own distinct biota (e.g. the United States) or because individual forest types are highly diverse in species (e.g. some lowland tropical moist forest). The former characteristic is generally related to the size of the country, the latter not necessarily so. Countries with very high forest diversity usually combine the two characteristics.

Forest ecosystem diversity. Ecosystem diversity in forests is important not only in its own right but also as a predictor of species diversity. However, assessing ecosystem diversity requires meaningful systems of classification of forests, and these are increasingly difficult to generate as the geographical scale increases. Scientists may develop schemes that are useful in local or national contexts for expressing differences between forest ecosystems, based on combinations of physiognomy, phenology and floristics. However, harmonizing several different schemes to produce overviews on a continental or global scale is a difficult task which may entail so much simplification as to eliminate any real utility in expressing ecosystem diversity.

For example, the harmonization of forest cover data from 70 different sources to produce the first global map of closed forest distribution, *The WWF world forest map* (WCMC, 1996a), resulted in a simplistic classification with only five classes (temperate needleleaf, temperate broadleaf and mixed, tropical moist, tropical dry and mangrove forests), reflecting the difficulty of combining the many much more detailed classification systems used throughout the world. A more detailed classification involving 25 classes which was used by WCMC and CIFOR in the production of a subsequent digital data set on global forest cover (Iremonger, Ravilious and Quinton, 1997) is more meaningful in biodiversity terms but is still very simplistic.

Remote sensing data are potentially a powerful source of information on forest ecosystem distribution and diversity. However, many regionally focused programmes, such as TREES and Pathfinder,[7] have been forced to use simplified classification schemes because of the analytical complexity of extracting more detail from satellite data. Global satellite-derived land cover data are available from the EROS Data Center in preliminary form. This data set helps to provide a consistent global view of forests' location and information about their diversity according to a number of classification schemes ranging from the simple to the complex, based on phenology as well as spectral response. The greatest advance in examining forest ecosystem diversity on a broad geographic scale is likely to come through the combination of remote sensing data with ecoregional classifications as envisaged in FRA 2000.

Forest genetic resources. FAO, the International Plant Genetic Resources Institute (IPGRI) and the International Centre for Research in Agroforestry (ICRAF) are facilitating regional initiatives to assess the status of genetic resources of forest trees at species and provenance levels; to identify gaps or overlaps in protection and development efforts; and to elaborate action plans for the conservation, management, sustainable utilization and enhancement of genetic resources of priority species. Regional syntheses will be prepared from data provided by the countries.

At the global level, FAO is developing the Global Information System on Forest Genetic Resources (REFORGEN) for monitoring the status of genetic

---

[5] See WWF-US's Web site: http://www.worldwildlife.org.
[6] For more information, see Conservation International's Web site: http://www.conservation.org.
[7] Pathfinder is the Humid Tropical Forest Inventory Project, a collaborative research effort of the University of New Hampshire, the University of Maryland and the National Aeronautics and Space Administration, United States.

resources of woody species at the national level (see Box 5). This is the only global effort that specifically targets forest genetic resources.

A global perspective on biological diversity. Assembling a global perspective on biological diversity is a slow and complex task. Major advances have been made in compiling global data sets on current and original forest cover, but a global vegetation classification is still lacking. Other advances at the global level have come from the identification of areas important for species diversity in key groups, but congruence of diversity patterns between groups has been addressed only at the regional level, and even then not thoroughly.

The increasing volumes of data involved in the analysis and management of global forest biological diversity require new approaches to data management. National capacities to gather and manage data to generate useful information – both for national use and to contribute to understanding of the global picture – will need to be built. Recent experiences show that the greatest challenges are organizational, not technological. By focusing on the processes involved in creating environmental information, rather than concentrating on data, international efforts such as the Biodiversity Data Management Project (UNEP/WCMC, 1996) are making useful contributions.

In addition to information from national sources, international research and other efforts can make a useful contribution to improvement of the global biological diversity information base. Likely sources of such improvement in the near future include:

- FRA 2000, which will include a number of parameters related to biological diversity;
- results from national reporting efforts within the framework of the Convention on Biological Diversity, and national policies and plans on biological diversity elaborated within the framework of the convention;
- additional research aimed at examining the congruence of diversity patterns among taxonomic groups, species distributions and habitat requirements, the effects of different forest management practices on biological diversity and biological diversity's impact on forest functions and contribution to socio-economic development;
- REFORGEN for forest genetic resources.

Combining these and other initiatives should produce a more coherent global view of biological diversity in forests. However, the crucial step in maintaining biological diversity in forests is the translation of this knowledge into improved forest management and conservation practices throughout the world. ◆

BOX 5
REFORGEN: the FAO database on forest genetic resources

Forest genetic resources, including the diversity present in the thousands of useful tree species, constitute a resource of tremendous social, economic and environmental importance. Efforts to conserve and use forest genetic resources wisely are assisted by the Global Information System on Forest Genetic Resources (REFORGEN), which provides reliable and up-to-date information on the status of these resources and on related activities and programmes. FAO's Forestry Department, in close collaboration with governments and national institutes of many countries, has been working since 1993 on the development of the system, which now covers more than 1 600 species from 144 countries. It includes information on:

- institutions dealing with conservation and utilization of forest genetic resources in a given country;
- the main native and introduced tree species in the country and their major uses;
- tree species endangered at the species and/or population level;
- tree species managed for *in situ* conservation;
- *ex situ* conservation activities *in vivo* and *in vitro*;
- tree improvement programmes;
- availability of forest reproductive materials for conservation and research purposes.

Linkages to complementary data systems of other organizations are being developed. REFORGEN will soon be accessible through the Internet and on CD-ROM.

Status and trends in forest management

Sustainable forest management entails the balancing of the economic, environmental and social functions and values of forests for the benefit of present and future generations – a complex and challenging task in the face of the Earth's rapidly expanding population and increasing demands for forest products and services. However, a number of positive developments give cause for cautious optimism. Among these is the unprecedented level of attention, energy and commitment from thousands of formal and informal, global to local, governmental and non-governmental organizations around the world (see Part III for discussion of international initiatives).

While frequent reference is made in the literature to areas where forests are not being managed in a sustainable manner, the magnitude of the problem globally is not known at present. A previous study by FAO (1982) estimated that less than 5 percent of the world's tropical forest classified as productive was under intensive management in 1980.[8] FRA 2000 will assess both quantitative and qualitative data pertaining to sustainable forest management worldwide. Recent figures indicate that almost all European forests are managed,[9] i.e. have a formal or informal management plan for one or a combination of objectives (e.g. wood production, conservation of biological diversity, soil and water conservation, recreation) (Third Ministerial Conference on the Protection of Forests in Europe/ Ministry of Agriculture, Rural Development and Fisheries of Portugal, 1998a).

Sustainable management of forests implies the provision of a range of goods and services. While all forests, by their nature, are multipurpose, and while more emphasis is given to management for multiple functions, most managed forests have a primary management objective: production (of wood and/or non-wood products), protection (mainly for soil and water conservation) or conservation (of biological diversity, cultural heritage sites, etc.). The following discussions of natural forests and forest plantations focus on management where wood production is the primary objective, as issues related to timber harvesting are among the more contentious in the forest debate today. Social and environmental functions of forests are treated in a subsequent section.

MANAGEMENT OF NATURAL FORESTS

There is a global commitment to improve the management of forests. This improvement entails a rebalancing of forest management objectives, which is bringing about changes in the way forests are managed and, in some places, a reallocation of forest areas between different uses. This may result in reduced timber harvests in natural forests, but increased production of other goods and services. Implementing sustainable forest management may lead to reduced volumes harvested in the short term, yet there is an expectation that it will increase wood supply over the long term.

Natural forest management is currently being affected by changes in resource availability, developments in management objectives and practices, and institutional changes which have led to a more diverse set of forest managers.

Trends in resource availability and in wood supply

Area of natural forest available for wood supply. Nearly one-half (1 563 million hectares) of the area of natural forests worldwide (an estimated 3 221 million hectares) is considered to be available for wood supply under current legal and market conditions (i.e. there are no legal restrictions and forest cover is both economically and physically accessible) (FAO, 1998a)[10] (see Table 2). Concerns related to timber harvesting and the need to encourage the use of environmentally sound forest harvesting practices apply to this area. At least 42 percent (665 million hectares) of the area available for wood supply is estimated to be "undisturbed by humans"; three-quarters of this area is located in the Russian Federation.

An additional 365 million hectares of natural forest are currently unavailable for wood supply because of transport or infrastructure constraints but have potential for commercial timber harvesting should economic conditions change and/or government policies encourage commercial development of this resource. Countries will face important choices about how this land can best be used to support national sustainable development. Some of this forest will undoubtedly be cleared for other uses, some may be added to national protected area systems, and some will be managed for commercial timber production. Some will remain too remote, and thus uneconomic, for commercial forest activities. It will be important for countries to make careful policy decisions related to land use in these areas, and to encourage sustainable forest management, including environmentally sound forest harvesting practices and appropriate silvicultural treatments in areas that are eventually designated for timber production.

Changes in natural forest area available for wood supply. The area of natural forest that is currently available for wood production is diminishing because of deforestation and the designation of some forests as strict conservation areas. The Philippines, for instance, has recently banned all logging in "old growth and virgin forests" and placed such forests under the National Integrated Protected Area System. In China, a similar ban on timber harvesting in natural forests was imposed in July 1998. Proposals have been made to add about 60 percent of the State-owned natural forests (approximately 25 million hectares) to the country's protected area system, which would have an overall effect of reducing log production from the State forest estate by 43 percent (*The China Daily*, 7 May 1998; ITTO, 1998). In Suriname, 1.5 million hectares of natural forest (one-tenth of the country's total land area) were set aside as a Wilderness

TABLE 2
Natural forests – area available and unavailable for wood supply

Natural forest classification under the Global Fibre Supply Model	Area *(million ha)*
Area available for wood supply	**1 563**
Semi-natural[a]	898
Undisturbed by humans[b]	665
Area unavailable for wood supply	**1 657**
Legal restrictions[c]	290
Economic restrictions:	
Physical reasons[d]	256
Transport/infrastructure constraints[e]	365
Other[f]	746
Total area of natural forest	**3 221[g]**

Source: FAO, 1998a.
[a] Forest which is neither "forest undisturbed by humans" nor "plantation".
[b] Forest which shows natural forest dynamics, such as natural tree composition, occurrence of dead wood, natural age structure and natural regeneration processes, the area of which is large enough to maintain its natural characteristics and where there has been no significant human intervention or where the last significant human intervention was long enough to have allowed the natural species composition and processes to have become re-established.
[c] Forest with legal restrictions or restrictions resulting from other political decisions, which totally exclude or severely limit wood supply. Areas under IUCN management categories I and II are considered as protected from wood harvesting.
[d] Harvesting is economically unfeasible at the current wood price level because the terrain conditions require extraordinary equipment or time.
[e] Remote forests where transport distance for logs or products is prohibited by cost or where access is currently not provided. This area, however, can become accessible for wood supply if, for instance, the government or the industry invests in infrastructure.
[f] E.g. low growing stock volume, wood quality too low, no commercial species.
[g] Differences in totals are due to rounding.

[8] Defined in the study (FAO, 1982) as the controlled and strict application of regulations on timber harvesting combined with silvicultural treatments and protection against fire and diseases.
[9] Data exclude the Russian Federation.
[10] This study (FAO, 1998a) does not include "countries with very minor forest cover" or countries where political conditions made statistics impossible to collect. Thus the total amount of forest cover is slightly smaller than that reported in the Forest Resources Assessment 1995 interim study, as indicated in the *State of the World's Forests 1997* (FAO, 1997d). The estimates of area available for wood supply are based upon FAO's assessment of the best available information, bearing in mind market conditions and current government policy. Regional meetings were held to discuss the statistics in Asia, Africa, Latin America and North America. Figures for Europe were obtained from the ECE/FAO fifth European Timber Trends Study (Joint ECE/FAO Agriculture and Timber Division, 1996). Figures for the Russian Federation were supplied by the International Institute of Applied Systems Analysis.

Nature Reserve in 1998. In April 1998, the Government of Brazil announced its intention to put 25 million hectares of rain forest under protected area status. In addition, Brazil, Cambodia, New Zealand, Sri Lanka, Thailand and the United States, among others, have recently either banned or severely restricted timber harvesting in primary forests.

Other recent developments which may affect the area of natural forest available for wood supply are the restitution of large forest areas to past owners in Eastern Europe; devolution and decentralization of responsibility for forest management (e.g. in the Philippines); and the designation of large areas of public forests for the use of native communities (e.g. in countries in the Amazon region, the Philippines and Canada). New owners of small areas of forests or woodlands in many developed countries are increasingly likely to use the forest for aesthetic or recreational purposes rather than for timber production. Changing ownership patterns, in which larger areas of natural forest are divided into smaller, individually owned forest units, may in some cases make timber production uneconomical. The overall effect is expected to be a reduction in the area of natural forest available for wood supply.

Changes in the types of forest available for wood supply. Timber harvesting is gradually shifting from forests undisturbed by humans to semi-natural forests,[11] plantations and trees outside forests. Factors driving this shift include deforestation, past timber harvesting and the recent inclusion of areas of natural forest that were previously available for timber production into countries' protected area systems. (See also the section on global trends in forest products, below.) The shift has already taken place in Europe, where an estimated 85 percent of forests are considered semi-natural (Third Ministerial Conference on the Protection of Forests in Europe / Ministry of Agriculture, Rural Development and Fisheries of Portugal, 1998a). The area of semi-natural forests and forest fallows on agricultural land is expected to increase, mainly in the developing world. As a result, interest in managing this resource for wood production has grown in recent years, and considerable effort has been devoted to identifying means by which this can be done sustainably.

The impact of sustainable forest management on future wood supply. More than 150 countries are participating in regional and ecoregional processes to establish criteria and indicators for sustainable forest management (see Part III). Member countries of the International Tropical Timber Organization (ITTO), which account for more than 80 percent of the world's tropical forests and more than 95 percent of the global tropical timber trade, have made a commitment to have their exports of tropical timber and tropical timber products come from sustainably managed sources by the year 2000. ITTO consumer countries have made a commitment to maintain or achieve sustainable management of their own forests by 2000. The World Bank and WWF have joined forces in a recently announced programme aiming for a target of 200 million hectares of certifiable forests by 2005. (Certification is discussed below under "Global trends in forest products".) A variety of other efforts, ranging from model forests to research efforts to NGO initiatives, are helping to promote sustainable forest management.

A study carried out in conjunction with the FAO Global Fibre Supply Model reviewed recent efforts to estimate how implementation of sustainable forest management might affect wood supply (FAO, 1997a). Much of the information available was based on simulation models and trial operations, and should be interpreted with caution. Nonetheless, the review indicated that implementing sustainable forest management would lead to reductions in harvest volume, particularly over the short term, and that management costs could be expected to rise on average between 5 and 25 percent.

There is, however, an expectation that long-term supply will increase through application of sustainable forest management. In the tropics,

[11] "Semi-natural" forests are natural forests where the impact of (prior) human intervention is still evident (i.e. forests that are neither "undisturbed by humans" nor plantations).

much of this increase is expected to be realized through improved harvesting methods leading to the maintenance of site productivity and prevention of damage to immature stems. In temperate forests the long-term increase is expected to be less pronounced because of the high level of productivity already reached, and increases may be captured only by intensified silviculture.[12]

Recent developments in natural forest management and silviculture

Many wood producing countries will continue to rely on natural forests as the main source of their wood supply, at least for the short term; only a few countries currently harvest the majority of their wood from forest plantations and trees outside forests.[13] The challenge, therefore, is to manage natural forests designated for wood production in such a way as to meet economic and socio-cultural needs, while maintaining and enhancing the forests' ecological and environmental functions.

Increasingly, the trend in many countries is towards multipurpose management of forest ecosystems. Consideration of long-term impacts of management interventions is made at the landscape, subnational and national levels. This trend has been reinforced by recent efforts to identify and apply criteria and indicators for sustainable forest management.

The emphasis on sustainable forest management has led to changes both in silvicultural systems and in timber harvesting practices. At the operational level, the persons responsible for harvesting operations are increasingly being tasked with pre- and postharvesting silvicultural actions as an integral part of the harvesting system in countries where this has not traditionally been the case.

Recent developments in silviculture. Changes in silvicultural systems are occurring in all types of forests – temperate, boreal, tropical and subtropical.

In the temperate zone, revised silvicultural guidelines are being developed by several countries and groups of countries as an outcome of recent initiatives related to identification and use of criteria and indicators for sustainable forest management. The Pan-European Operational Level Guidelines for Sustainable Forest Management,[14] for instance, addresses regeneration, choice of silvicultural systems, tending and harvesting, the use of pesticides and herbicides, protection of key biotopes, sensitive areas and sites of specific historical, cultural or spiritual significance, among others. The new United Kingdom Forestry Standard (DANI, 1998) is one example of national guidelines prepared in accordance with the Pan-European guidelines.

Concepts which reflect silvicultural management changes adopted in North American temperate and boreal forests in the late 1980s and 1990s include "sustainable forestry", "ecological integrity", "mimicking natural disturbances" and "ecosystem management". The corresponding silvicultural systems are identified with another set of terms such as "new forestry", "variable retention silvicultural system" and "improvement cutting". Codes of best management practices or forest practices codes have been developed for virtually all regions of Canada and the United States. These codes, which can be voluntary or prescribed in statutes and regulations, have had a significant impact on forest management in both countries.

In the tropics, there are clear indicators of gradual change towards silvicultural practices that better reflect the multiple goals of sustainable forest management. The CELOS management system,[15] which has been undergoing field trials for several years in Suriname, is being adopted in other parts of Latin America and in Asia. For example, Indonesia and Nepal have developed shelterwood/ enrichment planting in dipterocarps, and Malaysia and Uganda are moving to the Selective

[12] In some temperate countries – notably in Europe – where forests have been intensively managed for a long time, an increase in wood supply may depend on more intensive management of small, privately owned woodlands. The conversion of marginal agricultural areas to forests may also result in increased wood supply.

[13] E.g. Bangladesh, Chile, Japan, New Zealand, Pakistan, South Africa, Sri Lanka and Zambia.

[14] Endorsed at the Third Ministerial Conference on the Protection of Forests in Europe, Lisbon, Portugal, 1998.

[15] The CELOS harvesting and silvicultural system is a polycyclical selection system based on 20- to 25-year cutting intervals and a minimum harvest size of 35 cm diameter at breast height (DBH).

Management System.[16] Recent attention has also been paid to improved management of secondary forests, particularly forests that have developed on agricultural land either as a fallow (i.e. forest fallow) or following abandonment of the land.[17]

Efforts to improve forest harvesting practices. Various efforts are under way to develop methods of timber harvesting that maintain both the potential for future production and the environmental services that forests provide. At the international level, FAO and CIFOR have been working on codes and guidelines for environmentally sound forest harvesting. FAO published a Model Code of Forest Harvesting Practices in 1996 (FAO, 1996b). CIFOR recently developed guidelines for reduced-impact logging in dipterocarp forests in Indonesia (Sist, Dykstra and Fimbel, 1998). A regional Code of Practice for Forest Harvesting developed for the Asia and the Pacific region under the auspices of the Asia-Pacific Forestry Commission (APFC) was adopted by the member countries at APFC's seventeenth session in February 1998. National codes in line with the regional code are currently being developed by several countries in the region, and a number of training workshops on implementation have been held. Box 6 reports on some of the expected results of implementing the FAO Model Code of Forest Harvesting Practices and the guidelines developed by CIFOR.

New managers and new management roles: a focus on people's participation in forest management

Recent changes in forest policy and legislation reflect support for increased involvement of the private sector (forest owners, forest industries, NGOs and community-based organizations, indigenous people and the general public) and local government units in forest management. Privatization, devolution and decentralization, recognition of indigenous peoples' rights to ancestral land and participatory approaches to forest management have all led to changing roles of forest administrations and a new type of forest manager (see also Part II).

As a result of both internal and external pressures, stemming from the growing recognition over the past two decades that active stakeholder participation is a key element of sustainable forest management, a variety of approaches to increasing the participation of local communities in natural forest management have been developed. The past few years have seen a significant acceleration in the implementation of community-based forest management programmes and considerable improvement in the results as experience, both good and bad, has accumulated.

Asia has been at the forefront of participatory forest management. Joint forest management in India, whose roots date back to the early 1970s in West Bengal, was one of the first initiatives. Under joint forest management, communities are given rights to manage areas of forest, as detailed in a management plan, in return for which they receive a portion of the resulting benefits. Through Nepal's community forestry programme (see Box 7), user groups manage forests according to a management plan drawn up with and approved by the District Forest Officer. In the Philippines, some forest management functions have recently been devolved to local government. A community-based forest management strategy has been adopted as

[16] Similar to CELOS, the Selective Management System is a polycyclical system in which a lower volume of wood is removed at each cut, especially in the first cycle, in order to sustain the long-term harvestable volumes. The system is based on 25- to 30-year cutting intervals and a minimum harvest size of 35 cm DBH (see FAO, 1997l).

[17] Secondary forests have been defined in various ways. Some definitions include previously harvested primary forests, while others include only forests that develop on agricultural land, either as fallow or following abandonment of the land. Recent events and activities regarding secondary forests include a side meeting on tropical secondary forest at the eleventh World Forestry Congress; the FAO/Netherlands International Workshop on the Sustainable Development of Tropical Secondary Forests in Pucallpa, Peru, June 1997; the establishment of a Tropical Secondary Forest Network; and projects implemented by the Tropical Agriculture Research and Higher Education Center (CATIE) and the Organization of Tropical Studies (OTS/OET) in Costa Rica, the Center for International Forestry Research (CIFOR) in Indonesia and the International Cooperation Centre on Agrarian Research for Development (CIRAD-Forêt) in West and Central Africa. The importance of secondary forests and their management is also receiving high priority in the framework of the Amazon Cooperation Treaty.

BOX 6
Environmentally sound forest harvesting

A requirement for the widespread adoption of environmentally sound timber harvesting technologies is the demonstration that such operations can be economically feasible, environmentally sound and socially acceptable. In response to this need, case studies sponsored by private enterprises have been undertaken in tropical forests in the Congo (FAO, 1997b), Brazil (FAO, 1997c) and Indonesia (FAO, 1998c) to test some of the applications suggested in the FAO Model Code of Forest Harvesting Practices.

These studies have found that:

- environmentally sound timber harvesting is not necessarily more expensive than traditional timber harvesting methods;
- use of the environmentally sound practices reduced damage to standing stock as much as 60 percent (from 51.5 to 22.2 percent in the Brazil study) relative to traditional harvesting methods;
- appropriate preharvest planning reduced the area of forest roads, skid trails and landings considerably (from 20 to 4.5 percent in the study in Brazil);
- disturbance of the canopy was reduced from 25 to 11 percent;
- total timber loss was reduced by more than 50 percent (from 8.5 to 3.9 percent of utilizable stem volume).

These findings correlate well with the expectations for reduced-impact logging for lowland and hill dipterocarp forests in Indonesia as stated in the CIFOR guidelines (Sist, Dykstra and Fimbel, 1998). The endorsed practices were expected to:

- reduce disturbances to soil and residual vegetation by at least 50 percent in comparison with conventional logging operations where the guidelines are not applied;
- limit overall direct impacts to the forest to less than 25 percent;
- conserve wildlife and other forest resources, including non-wood forest products, threatened and endangered species, keystone plant resources and water;
- diminish direct logging costs by at least 15 percent;
- protect the long-term integrity and value of the permanent forest estate.

A critical requirement for the effective application of advanced technology capable of reducing environmental impacts associated with logging is a well-designed and continuous programme of training for the logging crews, their supervisors and the persons responsible for harvest planning.

Achievements in environmentally sound road construction using hydraulic excavators and advanced blasting techniques have been thoroughly documented by recent case studies in Austria (FAO, 1998b) and Bhutan (FAO, 1998d). Trials with central tyre inflation (CTI) on a timber hauling truck (which allows the driver to vary the tyre pressure while the truck is in motion) found that a road that normally would have been closed during a period of reduced bearing capacity (e.g. spring thaw) coped comfortably with 30 passes by a CTI-equipped truck. A conventional truck with normal tyre pressure would have caused deep ruts in the road in just six trips (SkogForsk, 1998).

the main approach to sustainable development of the country's public forest resources.

Other regions are also seeing participatory approaches to forest management. In Africa, for example, several countries have adopted community-based forest management as their main strategy for managing forest resources. Participatory forest management is also being extensively used in Latin America, where efforts to involve indigenous peoples have been particularly noteworthy. Collaborative management of protected areas has become the primary approach of international NGOs such as WWF and IUCN in projects in Asia, Africa and Latin America.

These initiatives have certain features in common. For example, the community gains secure rights to the use of resources on forest land, rather than ownership of the land itself. Another is that even though the responsibility for managing forest areas is either shared or shifted primarily to the user manager, the authority in most cases continues to lie with State forest administrations. Thus, if

BOX 7
Examples of community-based forest management

Nepal
Under community forestry, the government may assign local forest users the right to manage an area of State forest. Users form forest user groups, which are responsible for setting and enforcing forest management rules, designed to ensure that extraction of forest products is limited to sustainable levels. The rules are included in a forest operational plan, which must be approved by the District Forest Office. The user group can sell the forest products, including timber, at prices that it sets, and it may use the income for forestry and community development activities.

The Gambia
Establishing community ownership of forest land has been an important strategy in promoting sustainable forest management practices in the Gambia. Local inhabitants previously had little incentive to conserve public forest resources and often overexploited resources on State forest lands. Recognition that access to forest ownership serves as motivation for preserving forests has led to the development of community forestry activities within the Gambia. The Gambian-German Forestry Project has sought collaboration between local forest committees and the Forestry Department in developing local management plans. These plans, known as Preliminary Community Forest Management Agreements, provide for community-based forest management over a three-year period. If the community displays an ability to manage the forest well, it is granted forest resources ownership rights. Communities are entitled to keep the benefits derived from the forest in exchange for managing the forest according to the established plan. Over 300 villages are currently involved in this project and numerous others also wish to participate.

Guatemala
In 1994, the Guatemalan Parliament approved a law enabling local communities to apply for forest concessions in the buffer zones of the Maya biosphere reserve, a protected area in northern Guatemala. A community must legally register its organization or association, apply for use of a specific area, develop a forest management plan for 30 or 40 years and elaborate a one-year operational plan. All these documents are presented to the National Council of Protected Areas (Consejo Nacional de Areas Protegidas – CONAP). CONAP technicians help with the development of the management plans and an NGO advises the community on both technical and legal aspects. The community organization agrees to pay CONAP 1 percent of the revenue generated by the forest. The community contracts private companies to harvest timber and non-wood forest products and monitors the forest management. To date, approximately 92 000 ha have been given over to such local concessions.

conflicts arise between the State agency and the user group, the State agency usually has the authority to resolve the dispute. In general, government support of local forest management is increasing, partly as a consequence of the positive results being obtained, but also because of diminishing financial and human resources at the central level.

Although participatory forest management has been implemented over a relatively short period, information is emerging which indicates that these approaches are having a positive effect on the condition of forest resources. This is partly because the direct accrual of benefits to local communities gives them greater incentive to manage and conserve forest resources actively. Several studies have been made of the benefits of shared forest management. For example, the Overseas Development Agency of the United Kingdom (ODA, 1996) concluded from the review of its projects in community-based forestry that several had positive environmental impacts, benefiting water supply, soil erosion control and biological diversity. The review also found that the projects had social and economic benefits, including strengthened local institutional capacity, improved relations among stakeholders (particularly communities and forest departments), increased product flows, greater product diversity and better access to markets.

A major challenge to be met in the future is to improve links between forest policy and ground-level implementation. Policy formulation needs to incorporate the experience of local-level forest users and managers, both traditional and non-traditional, and forest policies need to be better understood, accepted and effectively implemented at the local level.

FOREST PLANTATIONS

Forest plantations have attracted both considerable positive attention and criticism in recent years. Plantations will be increasingly important sources of industrial wood in the future, thus potentially allowing reduced levels of timber harvesting in natural forests. Some groups, however, oppose them as unsustainable monocultures unable to provide the multiple goods and services available from natural forests. There is concern that some natural tropical forests are being cleared and replaced with forest plantations. A closer look can shed some light on questions about the potential of plantations to meet demands for industrial wood products and thus their indirect role in conserving natural forest resources.

The area of plantations

The area of forest plantations in the world has been increasing for the past two decades, and this trend is expected to continue (see section on global trends in forest products). Viet Nam, for example, recently announced plans for the rehabilitation of 5 million hectares of forest land, of which about 3 million hectares would be forest plantations. Other countries with continuing afforestation programmes include (among others) Argentina, Brazil, Chile, China, India, Indonesia, Morocco, Thailand and Uruguay.

Several developed countries in the temperate and boreal regions have forests of native species that are largely regenerated naturally, but with supplementary planting as necessary. The difficulty of distinguishing forest plantations from natural forests – or rather, the fact that a hybrid of the two prevails in many countries, particularly in Europe, has made it impossible to get exact figures on the area of forest plantations in the developed world. An estimate, however, is that there are over 60 million hectares of forest plantations, of which about 29 million hectares are in European countries (including over 17 million hectares in the Russian Federation), over 13 million hectares in the United States, over 10 million hectares in Japan (Third Ministerial Conference on the Protection of Forests in Europe / Ministry of Agriculture, Rural Development and Fisheries of Portugal, 1998b),[18] 1.5 million hectares in New Zealand (New Zealand Ministry of Forestry, 1997)[19] and 1 million hectares in Australia (Bureau of Resource Sciences, 1997).[20] Forest plantations make up a large proportion of some countries' total forest cover (e.g. 44 percent in Japan and 19 percent in New Zealand). Not all of these plantations have been established for wood supply, however; many have been planted for protection functions, such as soil and water conservation, slope stabilization and wind protection.

There is a clearer distinction between natural forest and forest plantations in developing countries.

Table 3 shows the regional totals derived from a recent FAO review of the literature of forest plantations established for wood supply. Almost 75 percent of forest plantations are located in the Asia and the Pacific region (with 21 million hectares in China and 20 million hectares in India), while about 15 percent are in Latin America and 10 percent are in Africa. The reported annual afforestation rate in the tropics and subtropics in 1995 was about 3 million hectares per year.

An estimated 57 percent of the plantation area is planted with hardwood species and 43 percent with softwoods. Various species of pines make up the majority (61 percent) of the softwoods. Eucalypts comprise the largest area of hardwood plantations planted for industrial use (30 percent), followed by acacias (12 percent) and teak (about 7 percent). Short-rotation plantations of hardwood species have been grown for many years, but until recently there has been little interest in growing valuable

[18] Figure for 1997.
[19] Figure for 1996.
[20] Figure for 1994.

TABLE 3

Reported forest plantation areas and annual establishment rate in developing countries in 1995

(thousand ha)[a]

Region	Reported areas[b]			Estimated net areas	Area established per year
	Industrial	Non-industrial	Total		
Africa	3 787	3 025	6 812	5 861	288
Asia and Oceania	31 781	21 216	52 997	40 471	2 330
Latin America	7 826	2 134	9 960	8 898	401
Total	**43 394**	**26 375**	**69 769**	**55 230**	**3 019**

[a] The figures incorporate the latest revisions of area figures arising from recent inventories. They refer to forest plantations established for wood supply. They do not include plantation areas established for protection purposes or for non-wood forest products. They are thus considerably smaller than, and are not directly comparable with, the figures published in SOFO 1997 for base year 1995.

[b] "Reported areas" refer to gross areas derived from various published sources. The "estimated net areas" are derived from the reported areas through the application of reduction coefficients to allow for poor survival or other losses, based on inventory results where available and on expert opinion where not.

hardwood species, such as teak, because of their slow growth and delayed economic returns. Because of the prospect of reduced supplies of high-quality hardwood logs from natural forests, however, a number of countries – Costa Rica, Ghana, India and Malaysia among them – are now investing in valuable hardwood species, especially teak.[21]

Utilization of wood and fibre from species not traditionally considered forest crops – in particular rubber, coconut and oil-palm – has continued to increase. Coconut and oil-palm stems and the branches of rubberwood are used in various forms of reconstituted wood, and rubberwood stems are used for sawnwood. About 80 percent of the furniture manufactured in Malaysia is made from rubberwood; this industry was valued at approximately US$750 million in 1997. Furniture is also manufactured from rubberwood in Thailand, and this industry is being developed in China, India, Indonesia and Viet Nam. Rubberwood has in fact become so valuable that the Rubber Research Institute of Malaysia is now breeding dual-purpose latex / timber clones, which have recently been released for commercial planting (Forest Research Institute of Malaysia, personal communication, 1998).

The total plantation area of these three "non-forest" species was reported as more than 26 million hectares in 1995 (see Table 4), as compared with 14 million hectares reported for 1990 (FAO, 1997d). While much of this apparent increase is actually the result of improved data collection, there is good evidence that the plantation area of oil-palm is increasing significantly and that of rubber is increasing gradually, although that of coconut is decreasing. Most of the plantations are located in Asia and Oceania, but they are concentrated in a few countries. The majority of the coconut plantations are in Indonesia and the Philippines, those of rubber in Indonesia, Thailand and Malaysia and those of oil-palm in Malaysia and India. Wood and fibre are not harvested from all these plantations, but area figures illustrate their production potential.

Plantations' contribution to wood supply

The area of forest plantations established for industrial wood supply can help compensate for an anticipated reduction in production from natural forests because of deforestation or increased areas

[21] There is a risk, however, that a teak controversy may develop, stemming from environmental concerns, such as soil erosion associated with dense, pure plantations of teak planted on slopes, and fears of grower disappointment should the promised returns fail to materialize because of market conditions.

TABLE 4
Reported plantation areas of the major "non-forest" species in tropical and subtropical countries in 1995 (thousand ha)

Region	Rubber	Coconut	Oil-palm	Total
Latin America	238	269	265	772
Africa	529	461	922	1 912
Asia and Oceania	8 718	10 546	4 587	23 851
Total	**9 485**	**11 276**	**5 774**	**26 535**

being set aside for conservation or other reasons. For example, over 25 000 ha of high-yielding hybrid poplar plantations were established in the northwestern United States between 1992 and 1997 in response to both increased demand for poplar wood for oriented strand board and decreased supply from public forests (USDA Forest Service, 1996). There is also a perception that the establishment of a sufficient area of forest plantations may reduce logging pressure on natural forests by providing alternative sources of wood supply. One of the three objectives of the forest plantations programme of the Seventh Malaysia Plan, for example, is to reduce pressure on natural forests through higher timber outputs from concentrated plantation forests. A similar strategy has been applied in New Zealand (see Box 8).

No global figures are available for current output of timber from forest plantations, but the potential annual growth of industrial wood from forest plantations in developing countries has been estimated at about 5 percent of the increment of natural forests in 1995 (FAO, 1998a). In some countries, however, plantation production already makes up a significant portion of the industrial wood supply. For example, 99 percent of New Zealand's industrial roundwood in 1997 was grown in plantations, 84 percent of Chile's, 62 percent of Brazil's and 50 percent of Zambia's.

Projections of the future contribution of forest plantations to wood supply are based on various assumptions, the main one being the rate of afforestation. Assuming today's rates of deforestation and afforestation, it has been estimated that by 2010 the potential increment from forest plantations will be about 40 percent of that from natural forests in Asia, Oceania and Latin America and about 15 percent in Africa (FAO, 1998a).

Gains in productivity of forest plantations continue to be made through improved management, tree breeding and tree improvement. Simple selection, particularly of provenances, may give results nearly as dramatic as and often more resilient to external influences than intensive tree breeding. Recent advances in biotechnology are increasingly being applied to the forestry sector, resulting in improved yields. For example, researchers at Aracruz Florestal, the plantation company supplying a Brazilian pulp and paper company, have developed synthetic interspecific hybrids of *Eucalyptus grandis* × *Eucalyptus urophylla* which yield over 70 m^3 per hectare per year in commercial plantations on optimal sites, compared with the 45 m^3 per hectare per year previously yielded by natural hybrids. Breeding programmes can also achieve increases in fibre density or in the ratio of usable to unusable fibre, which may bring greater gain than increased yields. As illustrated by the rubberwood example, breeding can help make adjustments to adapt to new market demands. Finally, breeding programmes can be oriented towards minimizing risk to plantations from biological threats.

Current environmental, social and policy issues related to forest plantations

Many of the environmental and social issues surrounding forest plantations are related to the way they are established – blocks of trees of the

BOX 8

Substituting wood from natural forests with that of forest plantations: the case of New Zealand

Prior to human settlement, forests covered almost 80 percent of New Zealand's land area. Much of this forest was subsequently cleared for grazing. Today, natural closed forests constitute only about 23 percent of the country's land area. Since the 1920s, New Zealand has progressively developed a significant plantation forest estate, largely in response to the recognition that without a concerted effort to establish supplementary wood supplies, the natural forests would be rapidly depleted. Forest plantations now cover 5 percent of the land area (1.5 million hectares) and produce 99 percent of New Zealand's industrial wood harvest.

Three-quarters of the natural forest is State owned and protected in national parks and reserves. Of the remaining 25 percent in private hands, four-fifths is inaccessible or has protected status. This leaves only 5 percent of the natural forest (approximately 300 000 ha) effectively available for wood production. In general, this area may be harvested only under a government-approved sustainable management plan. (Exceptions are forests subject to the West Coast Forest Accord or the South Island Landless Maoris Act of 1906.)

The New Zealand experience shows that exchanging wood production from natural forests with plantations is not a matter of straightforward substitution, nor is it simple. Some of the considerations are as follows:

- Substitution is not immediate. Harvesting in natural forests will need to continue until plantation wood supplies come on stream, if production levels are to be maintained.
- Technical research and market development is often needed to help in the change-over from products of the natural forest to plantation wood. The Government of New Zealand initially invested in substantial plantation wood processing facilities to prove the efficacy of processing plantation timbers and to overcome some of the early technical difficulties.
- It will generally be much easier to impose harvesting restrictions on State-owned natural forest land than on privately owned land. Landowners may demand compensation for infringements on property rights and may be concerned about the costs of rigorous requirements for sustainable natural forest management.
- Forest policy for the management of natural forests may be contentious for many years. In New Zealand, management of natural forests harvested under the West Coast Forest Accord remains a source of major controversy. Tension exists between the relatively small industry still processing natural forest timbers and environmental groups and communities. Meanwhile, forest plantation practices remain subject to environmental criticism and stricture.
- The success of this strategy depends in part on the species planted and the end uses required. Species providing sawlogs or veneer logs not only are generally slower growing but may be more difficult to raise in plantations. In New Zealand's case this has not been an issue because the mainstay of the plantation programme is a utility-class timber species, radiata pine, which is relatively easy to grow in plantations and grows quickly enough to yield commercially acceptable rates of return.

Sources: FAO, 1997e; New Zealand Ministry of Forestry, 1996.

same age and same species planted in straight lines with regular spacing. Criticism of monoculture has increased interest in using more than one species in plantations. The potential benefits of mixed-species plantations include reduced risk of disease and pest outbreaks; improved nutrient cycling in forest soils; reduced risk of fire damage when suitable fire-resistant species are used; increased habitat diversity for native plant and animal species; increased market security through species and product diversification; and improved visual and amenity characteristics. However, apart from a few examples, such as the longstanding practice of using nurse species in mahogany plantations in West Africa, Central America and Oceania for shade and protection against attack by the *Hypsipyla* insect, both experience and quantitative information on the establishment and management of such mixed-species plantations are limited.

The availability of land for forest plantations is of growing concern. Private enterprise is increasingly conscious of environmental and social issues related to the establishment of large blocks of trees. The clearing of forest for plantations is reported to have ceased in Chile, Argentina and Brazil, although it is still continuing in Indonesia and some other countries. Large blocks of unencumbered land, even of low fertility, are increasingly difficult to find, particularly in Asia where the area under forest plantations has expanded most rapidly. This paucity of land has led to the growing of trees for industrial roundwood production on land outside forests, such as farmland, where trees are either planted as small blocks or integrated with agriculture in agroforestry systems. In Pakistan, for example, 45 percent of respondents to a survey on farm trees grew trees for timber production on irrigated farmland, which made a significant contribution to wood supplies (Leach, 1993).

"Outgrower" schemes encourage farmers to raise trees on their own land for sale to processing companies. There are many such schemes, for example *Populus deltoides* grown for peeler logs in India, *Albizzia falcataria* grown for pulpwood in the Philippines, and *Acacia mearnsii* grown for tan bark and eucalypts for pulp in South Africa. Under these schemes, outgrowers usually receive loans, improved planting stock, technical advice and a guaranteed price for the product.

Contract reforestation schemes allot plots of public land to farmers or other individuals on which they grow trees in return for part of the profit received from the sale of the product. Such schemes are found mainly in Asia, for example in Thailand, Indonesia and Viet Nam, supported to a great extent through loans provided by the Asian Development Bank.

The involvement of the private sector, from large-scale commercial industries to small-scale outgrowers, in the establishment of forest plantations throughout the world is of considerable and increasing importance. In most countries various incentives in the form of public subsidies are given for the establishment or management of forest plantations.

Lack of finances, however, may be less of a constraint to forest plantation investment than insecure tenure and lack of suitable lands. In a regional study of Asia and the Pacific (Chipeta, 1996), a sample of countries, donors and international organizations identified the main constraints for forestry investment (not only in forest plantations) to be inappropriate policies, poor institutional capacity and difficult bureaucratic procedures. Key problems faced by many developing countries in the region were weak business orientation of forest administrations; bureaucratic delays; and unsuitable tenure policies, laws or practices. The respondents noted that farmers and rural communities, who have an important investment potential in forest plantations, suffered from interim cash flow problems before harvesting the tree crop and lacked access to credit. The future development of plantation programmes around the world will depend upon countries' ability to resolve these problems. In addition, if plantation areas increase as expected, efforts to ensure their sustainable management will be given more emphasis. The maintenance of productivity and plantation management in general are likely in future to be measured against criteria and indicators of sustainability. ◆

Environmental and social services of forests

The environmental and social services of forests and trees include, among others, the conservation of biological diversity, carbon storage and sequestration for mitigation of global climate change, soil and water conservation, provision of employment and recreational opportunities, enhancement of agricultural production systems, improvement of urban and peri-urban living conditions and protection of natural and cultural heritage. These services have received increasing emphasis and, in some cases, global legal commitment since the United Nations Conference on Environment and Development (UNCED).

The following section focuses on environmental and social services provided by forests and trees in fragile and / or marginal ecosystems: in drylands, on mountains and in small island states. A second major topic discussed below is the role of forests in mitigating global climate change, a subject given greater prominence with the signing of the Kyoto Protocol of the Framework Convention on Climate Change (FCCC) in 1997.

THE ROLE OF FOREST RESOURCES IN FRAGILE ECOSYSTEMS

The three post-UNCED conventions – FCCC, the Convention on Biological Diversity (CBD) and the Convention to Combat Desertification (UNCCD) – are relevant to some of the environmental services of forests (see Part III for more detail). Concern for the conservation of biological diversity, which arose in the 1980s and has been strongly voiced by the public, NGOs and the media in many countries, has focused attention mainly on moist tropical forests (extending more recently to temperate and boreal forests) and has influenced much of the debate surrounding forest management. UNCED's Agenda 21 and the three conventions, by giving greater importance to a range of environmental and social functions of all forests, have been instrumental in widening the focus from moist tropical forests to include forests and trees in "fragile ecosystems", including drylands and mountains (covered by Agenda 21's Chapters 12 and 13, respectively). Small island ecosystems were not made the subject of a chapter of Agenda 21, but the Programme of Action for the Sustainable Development of Small Island Developing States, which was developed in 1994, provides a similar international framework for action.

Although forests in arid zones, in some mountain areas and on small islands generally have lower economic value in terms of timber resources than do humid lowland forests, they often have high environmental and social importance at the local level. These environmentally fragile areas tend to be isolated and economically marginalized. Under such conditions, people generally have a relatively high dependence on local forest resources for various goods.

Greater international emphasis on fragile ecosystems and on social and environmental aspects of the forest sector has resulted in an elevated awareness of the importance of trees outside forests (e.g. trees in agricultural lands, those scattered in the rural landscape and those located in urban and peri-urban environments). These resources have often been overlooked because forests have tended to dominate the attention of national forest administrations, have driven much of the international forest debate in the past and have captured the public attention in many countries.

Mountain forests: freshwater resources and other environmental benefits

Mountains occupy one-fifth of the world's land surface and are home to one-tenth of the world's population (Messerli and Ives, 1997). Large areas of the world's mountain and upland areas are covered by forest. These forests have local, regional and in some cases global value as sources of water supplies, centres of biological diversity, providers of a range of wood and non-wood products, sites for recreation and stabilizers of land against erosion.

Global concern over water resources – reflected in the choice of "Strategic approaches to freshwater management" as the theme for the sixth session of the UN Commission for Sustainable Development in July 1998 – stems from the fact that inadequate and unsafe water supplies are aggravating problems of ill health and food insecurity among increasing numbers of people worldwide. It is predicted that by 2025 as much as one-third of the world's population may suffer from water shortages (Lininger, Weingartner and Grosjean, 1998).

Mountain catchment areas have a critical role in hydrology. Mountains, which have higher levels of precipitation and (particularly cloud forests) capture atmospheric water more effectively than lowland areas, are the source of more than half of the world's freshwater and supply all the world's major rivers and many smaller ones. Mountains provide critical storage of freshwater in lakes, wetlands and reservoirs and in winter as ice and snow, which is later released to rivers and streams. Many streams would cease to flow altogether during the dry season if their headwaters did not supply this delayed flow. In semi-arid and arid regions over 90 percent of river flow comes from the mountains. Although the alpine catchment of the Rhine River occupies only 11 percent of the river basin, it supplies 31 percent of the annual flow and more than 50 percent of the summer flow (Price, 1998).

Mountain ecosystems are important globally as centres of biological diversity. The greatest diversity of vascular plant species is found in mountains, notably in Costa Rica, the tropical eastern Andes, the Atlantic forest of Brazil, the eastern Himalayan-Yunnan region, northern Borneo and Papua New Guinea (Barthlott, Lauer and Placke, 1996). Other important centres are found in arid subtropical mountains. Many of these areas are included in national protected area systems. Several are also biosphere reserves under the Man and the Biosphere Programme of the United Nations Educational, Scientific and Cultural Organization (UNESCO).

Major land cover and land use changes are occurring in mountains and highlands throughout the world. Mountain areas in many developing countries are characterized by a high rate of population increase and resulting land scarcity, poverty and natural resource degradation. The rate of deforestation and the degree of forest degradation of these mountain areas are considerable. While deforestation of tropical rain forests remains the focus of public attention, tropical upland forests had the highest rate of deforestation of any biome – 1.1 percent per year – between 1980 and 1990 (FAO, 1995). Rates of clearing are particularly high in Central America, East and Central Africa, Southeast Asia and the Andes. While deforestation in and of itself does not imply increased erosion, the risk is likely to be higher if it is followed (as it often is) by inappropriate land management (including poor cultivation practices), insufficient vegetative cover and / or lack of appropriate mechanical soil protection measures. Decreased tree and forest cover can also contribute to shortages of fuelwood and other wood and non-wood forest products.

The general trend in developed countries over the past several decades has largely been in the opposite direction: mountain communities are undergoing depopulation as people move to cities or elsewhere for better employment opportunities. In some places, permanent residents have been replaced with owners of second homes. Mountain areas abandoned by people and taken out of agriculture are often recolonized by trees and shrubs. In many places the management objectives of mountain forests are changing. Less emphasis is placed on production of wood and non-wood products for local use, and more on recreation and nature conservation. These forests have new management and protection needs.

Valuation of watershed protection and other environmental services can help illustrate the economic importance of mountain areas and lend support for financing conservation and sustainable development efforts in mountain areas. An environmental valuation study revealed that the Sierra Nevada ecosystem in the United States produces commodities and services valued at about US$2 200 million annually, of which 61 percent comprises water resources. Yet only minor

BOX 9
New steps to protect New York City's drinking water

The United States Department of Agriculture and the city of New York together will provide US$10.4 million to pay farmers to retire over 1 200 ha of highly eroded cropland and to establish 800 ha of forest to protect over 250 km of streams north of the city. The New York City water system is one of the largest in the world, serving more than 9 million people. It is supplied by a reservoir system covering more than 400 000 ha. The government plan has been launched in an effort to avoid having to build a water filtration plant for the city, which would cost up to US$8 000 million to build and $1 000 million a day to operate.

Source: Reuters News Service, 8 August 1998.

investment is made in the mountain ecosystem to ensure continued supply of these goods and services.

Reaching an equitable balance in the distribution of the costs and the benefits of watershed protection is not easy. Upland dwellers almost always bear most of the costs of protecting upland forests for the benefit of downstream water users. To help rectify this imbalance, a number of upstream-downstream partnerships and mechanisms to channel resources from lowland to upland areas are being developed in various places around the world (Preston, 1997). For example, the United States Department of Agriculture and the City of New York have recently announced a watershed protection programme which will provide funds for farmers to implement watershed protection measures in an effort to improve the quality of the New York City water supply (see Box 9). In Chile, individuals are granted perpetual, irreversible and freely tradable water use rights independent of landownership and use, based on which an active market for water rights has arisen. Other innovative mechanisms tried elsewhere include improved tenure rights and user fees.

Mountain issues have gained increased visibility and support in recent years through a range of international, regional and national initiatives.[22] Mountains are represented in the global agenda through Chapter 13 of Agenda 21, "Managing fragile ecosystems: sustainable mountain development". The United Nations Economic and Social Council (ECOSOC) recently adopted a resolution to proclaim the year 2002 as the International Year of the Mountains, a resolution expected to be approved by the United Nations General Assembly at the end of 1998. The fourth Conference of the Parties to the Convention on Biological Diversity in 1998 decided to include mountain ecosystems as a thematic issue for the seventh Conference of the Parties in 2001. At the regional level, many intergovernmental, NGO and technical meetings on sustainable mountain development have recently taken place. Despite these positive developments, the achievement of sustainable mountain development will ultimately depend on concerted efforts at the national and local levels. Several countries have been focusing to a greater extent on conservation and development needs in mountain ecosystems, taking a more integrated approach to mountain development and using participatory approaches in managing the natural resources. These approaches appear to be providing a sound foundation for future efforts to achieve sustainable mountain development.

Dryland forests and trees

While most dryland forests have relatively low potential for timber production, dryland trees and forests furnish a wide range of wood and non-wood products which are vital for local populations and provide many important environmental services. Because most of the products are collected for household use or for sale in local markets, their significance is often underestimated and insufficiently addressed by national policies and programmes. Dryland trees and forests provide fuelwood and small roundwood (e.g. poles, house frames, handles) and a range of non-wood

[22] The Mountain Forum, a recently formed global network consisting of international agencies, government representatives, research institutions, NGOs and individuals, maintains a Web site (http://www.mtnforum.org) which makes information available on new mountain initiatives, mountain issues and reference materials.

products, including foods, medicinal products, raw material for handicrafts, bushmeat and fodder. Among the many environmental services they provide, the most critical in many places are soil conservation (i.e. protection against erosion and maintenance of fertility), shelter against wind and shade.

Much of the early attention on dryland forests, dating from the early 1970s, was focused on fuelwood needs. Concerns about fuelwood supplies for both rural and urban populations led to the establishment of fuelwood plantations in many developing countries. Poor results, however, caused a subsequent shift in strategy for meeting fuelwood demands away from plantations and towards improved management of existing forest and tree resources (including encouragement of natural and assisted regeneration). Other more recent trends include greater emphasis on multiple use management for various wood and non-wood products throughout the world; agroforestry development, both crop- and animal-based systems (in Australia, China and India, for example); consolidation of participatory forest management models (e.g. Burkina Faso, the Gambia and Mali); and support for production and marketing cooperatives for fuelwood and poles (e.g. Burkina Faso, India, the Niger). These approaches have increasingly been institutionalized and a legal basis for participatory management of dryland resources has been strengthened in many countries. Several governments have been transferring responsibility for forest management to populations and private partners while maintaining an oversight role. Security of land and tree tenure and access to common property resources, including silvipastoral resources, have been recognized as critical issues.

FAO, the Swedish International Development Agency, the Swedish University for Agricultural Sciences and the International Cooperation Centre on Agrarian Research for Development (CIRAD-Forêt), in cooperation with many experts from Africa, Asia and Latin America, have documented the situation and latest developments in natural dryland forest management (FAO, 1997f). The publication reviews the biological and ecological conditions of natural dryland forests in Africa, Asia and Latin America and the processes affecting them. It examines the silviculture and management techniques that are, or can be, used to conserve and develop them. Various participatory models used for involving local communities in forest resources management and efforts to provide the communities with organizational and marketing support are also discussed.

The results of this study were discussed, along with other issues related to dryland conservation and forest management, at the International Expert Consultation on the Role of Forestry in Combating Desertification, a satellite meeting of the eleventh World Forestry Congress (see Part III). The meeting recognized the need to continue gathering basic information on the status and management of dryland forest resources, to adapt management approaches to changes brought about by democratization and decentralization and to provide support to help local communities manage and benefit from these forest resources.

Management of dryland forest and tree resources continues to receive significant attention at the national level and through international efforts, including the work of the Intergovernmental Panel on Forests (IPF) and the Intergovernmental Forum on Forests (IFF) as well as the Convention to Combat Desertification (see Part III). An expert meeting on rehabilitation of degraded forest ecosystems was held in Lisbon, Portugal in June 1996 in response to IPF's recommendation that dryland forest-related issues be further examined. The meeting made recommendations concerning the needs for a supportive policy framework, clear and simple technical packages, environmental planning and precautions and adoption of participatory approaches to resource management. The results of the meeting provided input to the IPF report *Fragile ecosystems affected by desertification and the impact of air-borne pollution on forests* (IPF, 1996) and were reflected in IPF's subsequent proposals for action. The second session of IFF (Geneva, Switzerland, 24 August to 4 September 1998) reviewed progress in implementing IPF's proposals for action, including those related to fragile ecosystems affected by desertification and drought and those regarding countries with low forest cover.

Participatory methods promoted by UNCCD and

the partnerships among populations, local and national governments, NGOs and the donor community provide important opportunities to ensure that ecological and social services in drylands are considered in efforts to combat desertification. Forest conservation and development, agroforestry and vegetative methods of soil and water conservation have been included in most of the national planning exercises to launch implementation of UNCCD. Strong recommendations have been made to coordinate UNCCD implementation and the recommendations of IPF regarding the restoration of degraded forest ecosystems and initiatives in countries of low forest cover. Expected future developments will certainly confirm further integration of forest activities in desertification control, strengthen the synergy between the IPF and IFF processes and the implementation of UNCCD and help reinforce links among various international frameworks dealing with natural resources management in dry areas.

Role of forests and trees in small island States

Attention to small island States has increased substantially since the early 1990s. Small island States were given an international political identity with the establishment of the Alliance of Small Island States (AOSIS) in 1991.[23] The Global Conference on the Sustainable Development of Small Island Developing States, held in Barbados in April 1994, resulted in the Barbados Programme of Action for the Sustainable Development of Small Island Developing States. A comprehensive review of the implementation of the Barbados Programme is currently under way, and the results are to be presented at a UN General Assembly Special Session in 1999. Other events in 1999 will include a donor conference with UNDP and AOSIS in February and a Special Conference at Ministerial Level for Small Island Developing States organized by FAO in March. The latter conference will focus on the development of a Plan of Action, as follow-up to UNCED and the World Food Summit, and on issues related to the World Trade Organization (WTO) negotiations on agriculture, forestry and fisheries.

While forest cover on the 52 small island States and dependent territories[24] is insignificant in global terms (representing less than 1 percent of the forest area of the world), forests and trees on these islands are extremely important for the well-being of the local inhabitants. In addition, forest resources on several islands have global significance for the conservation of biological diversity.

As a group, the small island States and territories are relatively well endowed with forests (see Table 5 for regional figures and Tables 2 and 3 of Annex 3 for country figures). In 1995 it was estimated that forests covered 34 percent of the total land area of these islands, compared with the world average of 26 percent. Within the group there is a great deal of variability; forest cover ranges from over 70 percent of land area in Solomon Islands and Vanuatu to under 10 percent in many of the smaller States and territories – particularly in Oceania and Africa – and less than 1 percent in Haiti. Although average forest cover is relatively high, the annual deforestation rate from 1990 to 1995 was almost three times higher than the world average (0.8 percent compared with 0.3 percent). The highest rates of deforestation between 1990 and 1995 were found in various Caribbean islands and in the Comoros. The main causes of deforestation included conversion of forested land for agricultural use and for infrastructure development. Some heavily forested countries in the South Pacific are experiencing significant forest degradation because of heavy exploitation of timber resources (FAO, 1997g). Deforestation and forest degradation affect not only the socio-economic well-being of local populations, but the environmental conditions on the islands and in surrounding marine ecosystems.

---

[23] AOSIS is a grouping within the G77 countries composed of 33 independent island nations and four low-lying coastal States, with five dependent territories as observers.

[24] Several definitions exist for the term "small island States". The discussion here refers to the 52 developing and developed small island States and dependent territories that have an upper limit of land area of 50 000 km^2. Most of the environmental and social services discussed in relation to these entities apply to small islands in general, including those that are part of a larger country.

TABLE 5

Land and forest area of small island States and dependent territories

Region	Land area (thousand ha)	Total forest 1995		Total forest 1990 (thousand ha)	Change in forest cover 1990-1995		
		(thousand ha)	(% of land area)		Total change 1990-1995 (thousand ha)	Annual change (thousand ha)	Annual change rate (%)
Africa[a]	1 251	218	17.4	190	28	6	2.8
Asia[b]	93	4	4.3	4	0	0	0
Oceania[c]	8 767	4 964	56.3	5 054	-90	-18	-0.4
Europe and Near East[d]	956	140	14.6	140	0	0	0
North and Central America[e]	11 878	2 583	21.7	2 856	-273	-55	-2.0
Total, small islands	**22 945**	**7 909**	**34.5**	**8 244**	**-335**	**-67**	**-0.8**
Total, world	**13 048 410**	**3 454 382**	**26.5**	**3 510 728**	**-56 346**	**-11 269**	**-0.3**

Note: See Tables 2 and 3 of Annex 3 for data on individual countries. The regional breakdown is consistent with FAO's WAICENT geographical categories.
[a] Cape Verde, Comoros, Mauritius, Réunion, Sao Tome and Principe, Saint Helena, Seychelles.
[b] Macau, Maldives, Singapore.
[c] American Samoa, Cook Islands, Federated States of Micronesia, Fiji, French Polynesia, Guam, Kiribati, Marshall Islands, Nauru, New Caledonia, Niue, Northern Mariana Islands, Palau, Samoa, Solomon Islands, Tuvalu, Tonga, Vanuatu.
[d] Cyprus, Malta.
[e] Antigua and Barbuda, Bahamas, Barbados, Bermuda, British Virgin Islands, Cayman Islands, Dominica, Dominican Republic, Grenada, Guadeloupe, Haiti, Jamaica, Martinique, Monserrat, Netherlands Antilles, Puerto Rico, Saint Kitts and Nevis, Saint Lucia, Saint Pierre and Miquelon, Saint Vincent and the Grenadines, Trinidad and Tobago, United States Virgin Islands.

Only a few small island States and dependent territories (Solomon Islands and Fiji, and to a lesser extent Jamaica and Vanuatu) produce and export industrial roundwood or processed wood products in significant quantities.[25] Many of the islands, however, have a high degree of dependence on forests for a variety of wood and non-wood products for household use. This is the case particularly of the more isolated island States, in Oceania for example, where physical and economic access to imported goods is limited.

The environmental functions of forests and trees in most of the small island States and dependent territories by far outweigh their production value. Small islands, though otherwise extremely diverse, all have a high ratio of coastline to land area and relatively short distances between the uplands and the coast. The strong dynamic between the land and the sea defines some of the most important environmental services of forests on these islands, including the following.

- Forests on small islands have an important role in protecting watersheds, maintaining good water supplies and protecting the marine environment. This function is particularly important for islands with strong topographic relief, such as several in the Caribbean. Good vegetative cover on steep slopes is important to prevent erosion and high sediment load in rivers which, when emptying into the sea, would smother coral reefs, sea grass beds and other near-shore environments.
- Mangroves and other tidal forests are highly productive ecosystems which are important

[25] Solomon Islands is the world's sixth largest exporter of tropical hardwood logs, and forestry accounts for more than 50 percent of its export revenues. Sandalwood has been a notable export from Vanuatu for nearly a century.

feeding, breeding and nursery grounds for numerous commercial fish and shellfish, including most commercial tropical shrimp. Coastal forests also act as sediment traps for upland runoff sediments, thus protecting sea grass beds, near-shore reefs and shipping lanes from siltation.

- Forests in small islands are extremely important for coastal protection against cyclones, hurricanes and strong winds combined with high rainfall and storm surges. Mangroves and other coastal forests provide some protection to the coastline, and trees can shelter agricultural land from the effects of salt spray.

Perhaps the most important global environmental service provided by forests on small islands is the conservation of biological diversity, both in the forest and in associated ecosystems such as coral reefs. In recognition of their important heritage, most small island States are signatory to the Convention on Biological Diversity and almost all of the Pacific nations are signatory to the Convention for the Protection of the Natural Resources and Environment of the South Pacific Region. Small islands, because of their size and their physical isolation from other land masses, generally have lower species diversity of plants and animals but a higher percentage of endemism than do continental masses. For example, in the Dominican Republic, Fiji, Haiti, Jamaica and Mauritius more than 30 percent of the higher plant species are endemic. As for bird species, Solomon Islands and Fiji have 24 percent and 20 percent endemism respectively. States with a high percentage of endemic mammals include Mauritius (50 percent), Solomon Islands (36 percent) and Fiji (25 percent) (WRI/UNEP/UNDP/World Bank, 1996). Many of these plant and animal species are found in forests or are dependent on them.

Many small island nations are economically highly dependent on tourism. Although the forests on these small islands are rarely the primary attraction for overseas visitors, they may contribute to the islands' touristic appeal. Islands such as Pohnpei in the Federated States of Micronesia, Dominica, Jamaica and Saint Lucia have made efforts to develop the tourism potential of their forest areas. The role of coastal forests in maintaining the health of coral reefs, which in turn protect beaches from sand erosion, is indirect but critical to the tourism industry in some countries, for example in the Caribbean and some Pacific areas.

The many and important roles of trees and forests on small islands call for a holistic and integrated approach to forest management which takes into account not only the direct benefits obtainable from the forests but also the indirect benefits and the links with associated natural ecosystems and other economic sectors.

FORESTS' ROLES IN MITIGATING GLOBAL CLIMATE CHANGE

The concentrations of greenhouse gases in the Earth's atmosphere have increased since the onset of the industrial revolution, largely because of human activities such as fossil fuel combustion and land use conversion. The concentration of carbon dioxide (CO_2), the main greenhouse gas emitted by human activities, increased by nearly 80 parts per million by volume (ppmv) between 1880 and 1994, after having fluctuated within a range of about 10 ppmv over the previous 1 000 years (Schimel *et al.*, 1996). The precise impact of increased greenhouse gas concentrations on global climate patterns remains difficult to predict. However, there has been an observable increase in the global average surface temperature, and the Intergovernmental Panel on Climate Change (IPCC) has concluded, "The balance of evidence suggests that there is a discernible human influence on climate" (Houghton *et al.*, 1996).

Forests can serve as reservoirs, sinks and sources of greenhouse gases and thus have a significant role in moderating the net flux of greenhouse gases between the land and the atmosphere. Forests act as reservoirs by storing carbon in biomass and soils. They are sinks of carbon when their area or productivity is increased, resulting in the uptake of atmospheric CO_2. Conversely, they are sources when the burning and decay of biomass and disturbance of soil result in emissions of CO_2 and other greenhouse gases. Net CO_2 emissions from changes in land use (primarily deforestation occurring mainly in tropical areas) currently contribute about 20 percent of global anthropogenic CO_2 emissions (Schimel *et al.*, 1996).

Various practices related to the forest sector can be grouped according to their roles in helping to slow the accumulation of CO_2 in the atmosphere (Brown *et al.*, 1996).

- ***Conservation management.*** Existing stocks of carbon in forests can be maintained through forest preservation, sustainable harvesting and increased productivity on existing agricultural land which can reduce the rate of deforestation and forest degradation and prevent associated CO_2 emissions.
- ***Storage management.*** Activities that increase carbon storage in forests and forest products include increasing the forest area, increasing the forest carbon stored per unit of area through silvicultural measures (e.g. longer rotations, increased tree stocking densities, reduced impact logging) and extending the time over which harvested wood remains in use. These activities lead to a net uptake of CO_2 from the atmosphere.
- ***Substitution management.*** Substituting fuelwood from sustainably managed forests for fossil fuels produces a CO_2 benefit when the emissions from biomass combustion are offset by biomass growth, and emissions from fossil fuel combustion are avoided. Substituting wood products for more energy-demanding products, such as steel and concrete, can reduce the CO_2 emissions from the product manufacturing industries.

The CO_2 benefits from these management strategies vary considerably in terms of their timing, magnitude and permanence. For example, the timing and magnitude of CO_2 benefits from increased forest production (i.e. storage management) depend on the rate and duration of biomass growth. In the case of both conservation and storage management projects, the permanence of the CO_2 benefits depends on the protection of carbon stocks against natural threats (e.g. forest fires, storms and diseases) and human threats (e.g. clearance of the forest and conversion to another land use). For example, the CO_2 benefits of forest conservation may be negated if the forest is eventually burned or overharvested, or if decreased wood harvesting there results in more harvesting elsewhere. In the case of substitution of fuelwood for fossil fuels, the CO_2 benefits occur at the time when fossil fuel combustion is avoided. These benefits are considered permanent, and fossil fuel substitution can be achieved repeatedly with continuing cycles of forest harvest. The CO_2 benefits from wood product substitution vary according to the type and quantity of material displaced, the useful lifetime of the product (including recycling) and the method of wood product disposal.

There are obvious trade-offs among these three types of forest management strategy. For example, maximizing production of wood for storage or substitution management is likely to result in lower standing stocks of forest carbon than conservation management. Conversely, maximum protection of forests can result in increased use of non-renewable fossil fuels and emission-intensive non-wood materials. Substitution management may be an optimal strategy in cases where biomass growth rates are high and biomass efficiently displaces high-emission fuels or products. In the case of primary forests in regions with inefficient harvesting and manufacturing processes, forest conservation may provide more CO_2 benefits than substitution management. It is important, however, to recognize that these management strategies need not be mutually exclusive. For example, afforestation in combination with subsequent harvest can offer CO_2 benefits when the harvested wood either displaces fossil fuel use or offsets demand for wood from mature forests with high standing biomass.

Using forests to mitigate CO_2 emissions will clearly require comprehensive accounting of the associated carbon sources and sinks over time and comprehensive analysis of the other environmental and socio-economic criteria that influence forest management choices.

The Kyoto Protocol

IPCC has estimated, with a medium level of confidence, that globally, carbon sequestration from reduced deforestation, forest regeneration and increased development of plantations and agroforestry between 1995 and 2050 could amount to 12 to 15 percent of fossil fuel carbon emissions over the same period (Brown *et al.*, 1996). However, the future use of forest management strategies to meet national greenhouse gas reduction goals may depend in large part on the implementation of the Kyoto Protocol to the Framework Convention on Climate Change (see also Part III).

Negotiated during the third Conference of the Parties to FCCC in December 1997, the Kyoto Protocol, if ratified, would provide legally binding limits to greenhouse gas emissions by Annex I Parties.[26] The Kyoto Protocol makes provision for anthropogenic land use change and forestry activities to be used to define and meet national emission reduction targets in the first commitment period (2008 to 2012). It requires that countries account for fossil fuel emissions. As a consequence, there is an incentive to substitute wood and other biomass for more energy-intensive materials and fossil fuels. The Kyoto Protocol also provides limited encouragement for storage management by accounting for verifiable changes in carbon stocks from afforestation and reforestation. It gives limited encouragement for conservation activities because countries must report deforestation. Articles 2, 3 and 6 of the Kyoto Protocol specifically deal with land use change and forestry.

- Under Article 2, each Party agrees to protect and enhance greenhouse gas sinks and reservoirs and to promote sustainable forest management practices, afforestation and reforestation. This article does not appear to restrict the scope of land use change and forestry activities that may be implemented to mitigate greenhouse gas emissions.
- According to Article 3.3, "The net changes in greenhouse gas emissions by sources and removals by sinks resulting from direct human-induced land use change and forestry activities, limited to afforestation, reforestation and deforestation since 1990, measured as verifiable changes in carbon stocks in each commitment period, shall be used to meet the commitments under this Article of each Party". This language restricts the scope of land use change and forestry activities that may be used to meet national emission reduction commitments.
- Article 3.4 states that the meeting of the Parties shall "decide on which additional human-induced activities [besides afforestation, reforestation and deforestation] related to changes in greenhouse gas emissions and removals by sinks in the agricultural soil and land use change and forestry categories shall be added to, or subtracted from, the assigned amount for Parties included in Annex I". This decision will be taken by the first meeting of the Parties after the Kyoto Protocol has entered into force, based on the findings of a special report that IPCC will prepare by the year 2000.
- Under Article 3.7, those Parties with net greenhouse gas emissions from land use change and forestry in 1990 are instructed to include those net emissions in their 1990 base-year emissions inventory. However, those Parties with net sequestration from land use change and forestry in 1990 are not to include this net sequestration in their 1990 inventory. Therefore, Parties may receive credit for reducing the rate of emissions from land use change activities, but they are not penalized if the rate of sequestration resulting from these activities is decreasing over time.
- Article 6 grants Annex I Parties the ability to transfer to, or acquire from, any other Annex I Party emission reduction units resulting from projects aimed at reducing anthropogenic emissions by sources or enhancing anthropogenic removals by sinks. These emission reduction units may be used to meet national commitments. The emission reductions or sink enhancements from these projects must be additional to those that would occur in the absence of these projects.

Article 12 provides for a Clean Development Mechanism (CDM) which enables Annex I and non-Annex I Parties to undertake greenhouse gas emission reduction projects in non-Annex I countries. These projects must also achieve other sustainable development goals. The certified emission reductions produced by these projects

[26] The Annex I Parties include developed countries and countries with economies in transition.

may be used by Annex I Parties to meet their national commitments. However, it is not specified which emission reduction activities are included in CDM or whether CDM includes projects to increase carbon sequestration. The emission reductions from these projects must be additional to those that would occur in the absence of these projects.

The language used in the Kyoto Protocol is undergoing intense scrutiny and many key issues have yet to be resolved. IPCC was asked to prepare a special report on carbon emissions from sources and removals by sinks from land use, land use change and forestry, to address issues such as the following.

- The *Revised 1996 IPCC guidelines for national greenhouse gas inventories* (IPCC, 1996) have not been designed for land use change and forestry inventories restricted to direct, human-induced activities since 1990, such as afforestation, reforestation and deforestation. Therefore, additional work may be needed to put in place modalities, rules and guidelines for implementing the land use change and forestry provisions of the Kyoto Protocol.
- The greenhouse gas benefits from land use change and forestry activities may involve a higher level of uncertainty and a lower level of permanence than the greenhouse gas benefits from activities in other sectors, particularly the energy sector. Therefore, some Parties have expressed concern about relying on emission reductions in the land use change and forestry sector.
- Further work is needed to clarify which land use change and forestry activities may qualify for inclusion in domestic greenhouse gas mitigation programmes, in the exchange of project-based greenhouse gas emission reductions among Annex I Parties and in CDM projects under the Kyoto Protocol.
- Key terms in the Kyoto Protocol, such as "reforestation" and "forest", are to be defined.

Prospects for forests to provide additional carbon benefits

It is clear that fossil fuel combustion accounts for the majority of CO_2 emissions globally, that it will continue to do so in the future and that the primary response for reducing net CO_2 emissions will have to come from controlling the release of fossil fuel carbon. Nevertheless, forest management decisions can be cost-effective means of reducing net emissions, either by diminishing the contribution of forests to global net emissions or by enhancing their importance as carbon sinks. The long-term contribution of forests for mitigating climate change will be to provide renewable materials and fuels that reduce reliance on fossil fuels, while still maintaining the role of forests as carbon reservoirs.

Analyses to date suggest that there may be some inexpensive opportunities to reduce current net CO_2 emissions through forest management activities and that these projects could have additional environmental benefits through protection of biological diversity, protection of watersheds, etc. If forestry projects can be integrated with the social, cultural and developmental needs of developed and developing countries, they could offer very attractive opportunities for providing carbon benefits, especially in the short term, while alternative energy system options are developed and the understanding of the costs of global climate change improves. The magnitude of benefits available through forest-sector activities will depend on the amount of land available, improvements of forest productivity and technical developments in the efficiency with which forest products are harvested and used. ◆

Global trends in forest products

To meet the needs for wood and non-wood products and at the same time fulfil demands for environmental and social services from forests is the challenge now facing the forest sector. Efforts to find an acceptable balance between production and protection and between use and conservation drive much of the debate surrounding the forest sector today.

The overall patterns of production and consumption of wood products are very different between developed and developing countries taken as a group. Developed countries account for 70 percent of the total world production and consumption of industrial wood products. Developing countries, on the other hand, produce and consume about 90 percent of the world's fuelwood and charcoal, which are the major household energy sources in many of these nations (see Figure 3). More fuelwood and charcoal are consumed each year in the world than industrial roundwood. Demand for fuelwood is expected to continue to increase at a rate of about 1.1 percent per year between now and 2010, while demand for industrial roundwood is expected to increase at a rate of about 1.7 percent per year over the same period. Factors that are expected to influence the ability to meet the increasing demand include increased sources of wood (e.g. plantations and trees outside forests), technological improvements in wood processing which will increase the efficiency of use of raw material, and increased use of recovered and non-wood fibre. Trade will continue to help balance wood deficits in one place with surpluses elsewhere.

While wood is the predominant commercial product from forests, increased attention is being paid to the actual and potential economic role of non-wood forest products (NWFPs). The importance of NWFPs to household and local economies, particularly among the poor in developing countries, is increasingly recognized, as is the need to consider them in forest management planning and in forest policy in many countries.

Wood energy is another area that is raising new interest. While fuelwood and charcoal remain significant sources of energy in developing countries, especially for domestic use, their potential to contribute to the modern energy sector as an alternative to fossil fuels is being investigated in several countries.

Issues related to trade in forest products and to trade and the environment continue to be highly visible in the global forest agenda. Trade issues were debated at length, although inconclusively, at the second session of IFF in September

FIGURE 3
Production and consumption of wood products in 1996

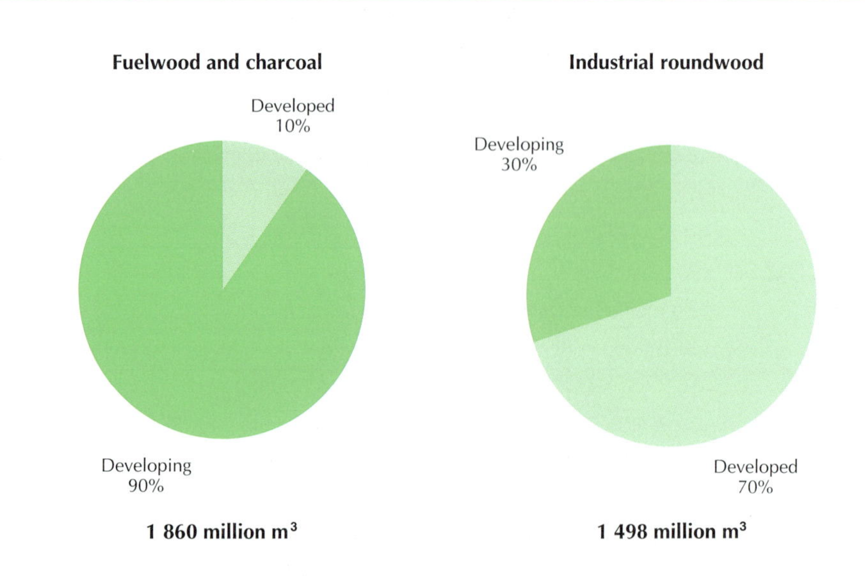

1998. A number of developments have occurred in the past two years regarding certification of forest products, one of the most complex and, in many cases, controversial issues in the forest sector. The Asian economic crisis has had the most unexpected and disruptive impact on forest products trade of any development in the 1997-1998 period.

NON-WOOD FOREST PRODUCTS: A LOOK AT MEDICINAL PLANTS

People have innumerable uses for the many plant and animal resources found in forests. Although several species have been domesticated and integrated into agricultural production schemes over the centuries, others, referred to as non-wood forest products, continue to be gathered from wild sources. In many parts of the world, NWFPs provide food (bushmeat, mushrooms, fruits, nuts, animal fodder), construction materials, fibres (bamboo, rattan, palm leaves), medicines and other health care products and goods of religious or spiritual significance. While the bulk of these products are gathered for household use or for sale in local markets, some enter national and international trade in significant quantities. NWFP production is often characterized by a large number of suppliers, each with a small scale of operation and a lack of industrial development. A global overview of major NWFPs, summarizing known information about their production status, value and trade and factors affecting their development was provided in the *State of the World's Forests 1997* (FAO, 1997d).

Various issues related to NWFPs are currently being discussed in regional and international fora. One issue relates to the need to ensure the conservation of forest-based biological diversity while still ensuring equitable access to forest resources (including NWFPs), particularly by local people. The development of appropriate and fair pricing methodologies for NWFPs (including royalties on intellectual property rights) is another need. Difficult access and/or insecure tenure rights to the resources and the absence of relevant market information, including fair market access, are among the key constraints faced by the subsector.

Medicinal plants are among the most valuable of the NWFPs. Not all, but most medicinal plants gathered from the wild come from forest lands. Their high value can provide an additional incentive for sustainable management of forest resources and for the conservation of specific habitats.

Use of medicinal plants

More than 10 000 plant species (of both forest and non-forest origins) are used for medicinal purposes, mainly as traditional medicines. WHO has estimated that 80 percent of people in developing countries rely on traditional medicines, which are mostly plant derived, for primary health care. The use of medicinal plants is by no means restricted to developing countries and traditional medicine, however; at least 25 percent of the drugs in the modern pharmacopoeia are derived from plants. Many others are synthetic analogues built on prototype compounds isolated from plants. The demand for medicinal plants is increasing in both developing and developed countries.

Collection and production of medicinal plants

The majority of plant material used for medicinal purposes comes from developing countries. Most of it is gathered from the wild, mainly from forests, for household use. Few medicinal species are cultivated, because the low price of material harvested from the wild still makes cultivation financially unattractive. It is expected that more of these species will be cultivated in the future, however, because sources of wild material are diminishing and cultivated material is far preferable to wild material for large-scale production of commercial drugs for reasons of efficiency and quality control. Standardization, whether for pure products, extracts or crude drugs, is critical and will become increasingly so as quality requirements continue to become more stringent throughout the world.

Wild sources of medicinal plants will continue to be important at least in the short term, and will remain so for much longer in developing areas of the world and for the poorer sectors of society. In addition, some species will be difficult to cultivate, and synthesis of some active ingredients will be problematic. It is therefore critical to ensure a combination of cultivation and / or sustainable wild harvesting of medicinal plants. Only the latter can serve, through sound management of these resources, to provide additional incentives to conserve the habitats with the broadest genetic variation.

Not only are millions of people dependent on these plants for health care, but harvesting medicinal plant material for commercial purposes may be one of the few opportunities for paid employment or for earning supplementary income in some remote rural areas. When a species becomes commercially interesting, however, control over the resource may be transferred to a concessionaire system (involving individuals, a company or a kind of "extractive reserve" community scheme) or a trading board, often depriving some local people of access to the resource, either for household use or as a source of income.

Policy and regulation of trade in medicinal plants

Most end users are unaware of the extent to which the expanding demand in medicinal plants is threatening the survival of several plant species. The prices paid to gatherers tend to be very low, and resources are frequently openly accessible or common property. As a result, commercial plant gatherers often mine the resources rather than managing them. The species most vulnerable to extinction are those that are in high demand, reproduce slowly and have specific habitat requirements and a limited distribution (e.g. *Warburgia salutaris* in eastern and southern Africa). There is also a clear relation between the part of the plant collected or the collection method used and the impact of harvesting. For example, heavy commercial exploitation of the bark of *Prunus africana,* which is used in an anti-cancer drug, has devastated populations of this tree throughout humid Africa.

Most countries have few or no regulations controlling the collection and trade of material from the wild, and exisiting national legislation may be insufficient or ineffective. In Bhutan, laws passed to ban collection of specific plants effectively increased their price and stimulated illegal harvesting, which virtually drove them to extinction locally. The introduction of harvesting restrictions or bans in one country can result in overharvesting in other exporting countries.

Most medicinal plants are traded in local or national markets; relatively few are traded internationally in significant volume. There are few reliable global or even national data on production and trade of wild harvested medicinal plants, and it is difficult to distinguish wild from cultivated sources in existing trade statistics on medicinal plant material. According to data compiled from the COMTRADE database of the United Nations Conference on Trade and Development (UNCTAD), the total value of medicinal plant exports in 1995 from approximately 100 countries amounted to US$880 million. Regionally, Asia leads in the supply and consumption of medicinal plants, followed by North America. Germany dominates the European trade in medicinal plants, importing plant material from over 100 countries and re-exporting one-third of it as finished products.

International trade in certain medicinal plants is monitored and regulated mainly through the Convention on International Trade in Endangered Species of Wild Fauna and Flora (CITES). Several national and international initiatives by governmental and / or non-governmental organizations are emerging to address the unsustainable rates of exploitation of many medicinal plant species. At the global level, among the most significant are the Medicinal Plant Specialist Group of IUCN and TRAFFIC, a wildlife trade monitoring programme of WWF and IUCN, which closely cooperates with the CITES Secretariat. TRAFFIC has recently launched a priority programme on medicinal wildlife trade, whose objectives for the 1997-2000 period are: to identify and predict possible threats posed to wild species by the medicinal trade and to indicate possible solutions; to examine existing local, national and international regulatory measures for wildlife medicinals and to seek modifications as required to assist in maintaining trade within sustainable levels; and to promote enforcement of and adherence to regulatory measures intended to conserve wild species in trade.

Clearly, achieving sustainable management of NWFPs in general, and medicinal plants in particular, will be a continuing challenge requiring concerted local, national and international action.

CONTRIBUTION OF WOODFUELS TO THE ENERGY SECTOR

Much of the wood harvested in the world each year is used for energy production. Of the estimated 3 350 million cubic metres of wood harvested in 1995, about 2 100 million cubic metres, or 63 percent, was used as woodfuel.[27] While in developed countries only 33 percent of the wood produced was used for energy purposes, in developing countries woodfuels accounted for 81 percent (91 percent in Africa, 82 percent in Asia and 70 percent in Latin America) of the wood harvested. The figures illustrate the importance of woodfuels in total wood production and their relevance for the forestry sector.[28]

Woodfuels remain significant sources of energy in developing countries, especially in the rural and domestic sectors. In recent years, however, they have been attracting attention as environmentally friendly modern energy carriers. Changes in energy policies have favoured the development of wood energy systems, and new biomass energy technologies are improving the economic feasibility of wood energy, particularly in countries that are heavily forested and have well-established wood processing industries. The environmental costs of fossil fuels are also making woodfuels more attractive.

Role of woodfuels in the forest and energy sectors

Woodfuels consist of fuelwood, charcoal and black liquor (a by-product of pulp and paper). Table 6 shows the consumption of different woodfuel types by region and provides a picture of forestry's contribution to the energy sector.

Woodfuels account for an estimated 7 percent of the world's total energy supply. In developing countries, however, where fuelwood is a major source of fuel for household use, the average share of woodfuels in total energy use is 15 percent. In 34 developing countries, fuelwood and charcoal supply more than 70 percent of national energy demand. Woodfuels constitute the major source of energy for most countries of sub-Saharan Africa, Central America and continental Southeast Asia. While fuelwood is the predominant form of wood energy used in rural areas of developing countries, charcoal remains a significant source of energy for many African, Asian and Latin American countries, mainly for urban households.

Woodfuels account for only 2 percent of the total energy used in developed countries. This figure, however, conceals great differences in use at the national and subnational levels. For example, in Europe, relatively small quantities of woodfuel are used in Belgium, Germany and the United Kingdom, while large amounts are consumed in the densely forested countries of Austria, Finland and Sweden. In Finland, wood energy supplies an estimated 17 percent of the national energy demand (FAO, 1997h). Black liquor accounts for a high proportion of the total woodfuels used in most developed countries; it is used by large pulp and paper industries to meet their needs for heat and power.

Recent developments in wood energy

Woodfuel consumption in developing countries has increased steadily along with growth in population, although the share of woodfuels in the national energy balance of these countries has progressively diminished as a result of the increased use of fossil fuels such as oil, coal and gas. Fossil fuels have continued to fulfil most of the increased demand for energy in most developed countries.

Actions by many countries to deregulate, liberalize and privatize energy markets over the past two decades have stimulated competition among energy suppliers and have presented new

---

[27] The difference between this figure and that provided in Figure 1 results from the inclusion of black liquor derived from wood in the calculation of the amount of wood used for energy generation.

[28] The source of these figures is FAO's Wood Energy Information System (FAO-WEIS). For additional details on the figures refer to the Web site: http://www-data.fao.org/waicent/faoinfo/forestry/energy/feforum.htm.

TABLE 6
Consumption of woodfuels and share of woodfuels in total energy use in 1995

Region	Woodfuels (million m³ equivalent)			Woodfuel's share in total energy use (%)
	Fuelwood	Charcoal	Black liquor	
Total, developing countries	**1 533**	**131**	**34**	**15**
Africa	445	72	3	35
Asia – developing	859	25	12	12
Oceania – developing	6	0	0	52
Latin America[a] and the Caribbean	223	34	19	12
Total, developed countries	**187**	**6**	**228**	**2**
Europe, Israel and Turkey	56	2	51	3
Former USSR[b]	32	0	8	1
Canada and United States	96	4	146	3
Australia, New Zealand and Japan	3	0	23	1
World	**1 720**	**137**	**262**	**7**

Source: FAO's Wood Energy Information System (FAO-WEIS). For additional details on these figures refer to the Web site http://www-data.fao.org/waicent/faoinfo/forestry/energy/feforum.htm.
[a] Including Mexico and all Central American and South American countries.
[b] Including Armenia, Azerbaijan, Belarus, Estonia, Georgia, Kazakhstan, Kyrgyzstan, Latvia, Lithuania, Republic of Moldova, Russian Federation, Tajikistan, Turkmenistan, Ukraine and Uzbekistan.

opportunities for other, non-fossil fuel, energy sources. Some countries have also raised taxes on fossil fuels, prompting decreased use of these fuels and in some cases increased use of other energy sources. In addition, the development and adoption of new technologies for the production, transport, handling and storage of woodfuels, more efficient combustion devices and improved systems for planning, management and organization of wood energy systems are helping to make woodfuels considerably more cost-competitive energy sources. In Sweden, for example, the price of energy generated with fossil fuels doubled between 1980 and 1997 because of increased taxes on fossil fuels, whereas that of wood energy remained stable (Thornqvist, 1998).

Several countries are changing their energy policies to encourage expanded use of wood energy. The European Commission's recently adopted "White Paper" (EC, 1997) gives special attention to bio-energy (including both woodfuels and agricultural energy crops) and constitutes a framework for the future development of renewable energy within the 16 European Union (EU) countries.

In 1994, the Finnish Government established objectives for the promotion of wood energy with the aim of increasing its use by 25 percent by 2005 (Nousiainen and Vesisenaho, 1998). Similar initiatives are being adopted in other countries. In Denmark, 50 percent of households are on district heating fuelled by biofuels. The Netherlands has launched a special investment programme for the promotion of power and heating plants using woody biomass as fuel.

The Canadian Forest Service began an initiative in 1995 to facilitate the introduction of bio-energy (mainly wood energy) for power generation in the

BOX 10
Bio-electricity in Nicaragua

Following the trend of most countries in Latin America, Nicaragua has passed an energy law with new rules for the participation of the private sector in the generation, transmission and distribution of electricity. Two sugar mills have taken advantage of this new law and are supplying about 27 megawatts equivalent (MWe) of electricity to the national grid using bagasse as a main fuel (during the sugar-cane season) and fuelwood (off-season) derived from more than 6 000 ha of eucalyptus plantations.

San Antonio is the largest sugar mill in Nicaragua. It has a contract with the Government of Nicaragua to supply bio-electricity using bagasse and fuelwood. The first eucalyptus plantations were established in the area of the sugar-cane plantations. Additional plantations have since been established on rented lands. Some of the fuelwood is being produced by local farmers, who have long-term contracts with the mill, which guarantees a price for the wood. By 1996, over 2 600 ha of fuelwood plantations had been established.

Victoria de Julio, the second largest sugar mill in Nicaragua, started operating in 1985. It has a generation capacity of 12 MWe electricity using bagasse and fuelwood and has ambitious expansion plans. The concept of electricity as a second product was integrated into the original design of the plant. The sugar-cane plantations have a circular configuration, as they are irrigated by a circular pivot system, and the eucalyptus is planted on the land between the sugar-cane plantations. There are now about 4 000 ha of eucalyptus plantations.

An economic study concluded that electricity generation in Nicaraguan sugar mills using these biofuels not only is technically feasible but appears to be competitive at a current selling price of US$0.057 per kilowatt-hour, against US$0.068 per kilowatt-hour for fuel oil. In addition, 73 percent of the income derived from the energy generated with eucalyptus remains in the Nicaraguan economy, compared with 14 to 30 percent in the case of electricity produced with fuel oil. In terms of employment generation, eucalyptus provides three times more jobs than does fuel oil.

Source: FAO, 1997i.

remote First Nations communities in northern Canada. The purpose was not only to provide more energy to these isolated areas but also to create employment and foster self-reliance of the indigenous communities.

Several successful bio-energy programmes have been established in developing countries (FAO, 1996e). Indonesia, Malaysia, the Philippines and Thailand have recently launched wood energy initiatives with the support of an economic cooperation programme between the European Commission and the Association of Southeast-Asian Nations (ASEAN), the EC-ASEAN COGEN Programme. A private company in Chile, Chilgener, has been supplying industries with energy generated from woody wastes since 1992. In Nicaragua, two sugar mills have diversified their sugar production with electricity generated from bagasse (during the sugar-cane production season) and fuelwood derived from eucalyptus plantations (when bagasse is unavailable) (see Box 10). Both mills sell this energy to the national power grid. A similar project is under way in Honduras.

Potential for wood energy development in the future

Woodfuels are expected to continue to have an important role for some time to come as a traditional source of energy in developing countries, particularly among low-income sectors of the population. Fuelwood and charcoal will also continue to be burned in limited quantities by households and small industries for specialized uses in developed countries. Recent policy changes and experiences with bio-energy programmes in several countries, however, indicate that woodfuels are becoming more attractive to countries as a modern, renewable energy source. They have

BOX 11

Status of information on woodfuels and efforts to improve it

There are major weaknesses and gaps in information on woodfuels, which make planning for current woodfuel production and use and modelling of future wood energy scenarios problematic. Lack of good data on woodfuel production and flows makes it difficult to address crucial issues on woodfuel supply, trade, use and substitution. The problems stem from methodologies and weak national capabilities in woodfuel data collection, compilation and presentation. The use of different terms, definitions and units makes it extremely hard to exchange and compare information available from different agencies at the national and international levels. Various regional and international organizations – including FAO, the International Energy Agency (IEA), the Asian Institute of Technology (AIT), the Latin American Energy Organization (OLADE) and the European Statistical Office (EUROSTAT) – collect and disseminate data on fuelwood and charcoal production and use. To enable comparison of national and international statistics on woodfuels, FAO, in conjunction with other relevant agencies, is now working on the development of unified wood energy terminology and conversion factors. This undertaking, combined with ongoing efforts to strengthen national capabilities in the collection and analysis of woodfuel data, should improve the quality and availability of information on woodfuels.

potential to become more competitive with fossil fuels in certain situations, both for economic reasons (as fast-growing fuelwood plantations, thinnings from timber plantations and residues from forest industries serve as locally available and inexpensive sources of energy) and for environmental reasons (related to efforts to mitigate global warming). The Framework Convention on Climate Change has recognized the potential role of woodfuels as part of a substitution strategy to reduce emissions of CO_2 from fossil fuels (see discussion of forests' role in mitigating global climate change, above). The Kyoto Protocol of the FCCC, if ratified, has the potential to play a catalytic role in the further development of wood energy.

In response to these recent developments, both the forest and energy sectors are likely to give increased consideration to woodfuels in the future. Improved planning will depend in part on a sound information base, but at present the global information on woodfuels is extremely weak (see Box 11). Continued efforts to improve information collection and analysis will be essential, as will be further assessment of the relative costs and benefits of woodfuels, fossil fuels and alternative sources of fuels as economic conditions and environmental commitments evolve.

TRADE IN FOREST PRODUCTS: CURRENT ISSUES AND INFLUENCES[29]

Growth in the production of tropical forest products has slowed over the past three to five years. The export of most products has followed a similar trend; export volumes of tropical logs, sawnwood and wood-based panels have decreased. Some of the trends reflect major structural changes that are unlikely to be reversed, while others are a response to normal short-term changes in market conditions. Factors influencing the changes include:

- increasing domestic consumption in important developing producer countries;
- reduced harvest levels because of both environmental concerns and export market conditions;
- a shift in exports from logs, and to a lesser extent sawnwood, towards higher-value products;
- most recently, marked reductions in demand in Asia, especially in Japan.

Asian economic crisis

The most dramatic of the short-term effects began in mid-1997 when a number of key Asian currencies

[29] All figures quoted in this section are from various issues of the *FAO Yearbook of Forest Products*, FAO, Rome.

suffered significant depreciation, signalling the beginning of what is now recognized as a major recession. In May 1998 the currencies of Malaysia, the Philippines, the Republic of Korea and Thailand were trading at about two-thirds their value of one year before. The Indonesian rupiah was then worth 25 percent of its May 1997 value. Major economic and structural problems became evident in most of these countries as liquidity tightened and major recessionary conditions took hold.

Key variables determining the effect of the economic crisis on trade in forest products included the dramatic changes in currency relativities; the extent to which Japan, the dominant importer, was affected by the recession; and the effects of the crisis in other regions. Until about September 1998, the recession was confined largely to the Asia and the Pacific region, but by then there were signs that the effects were beginning to extend to many other regions. The major impacts on forestry have been:

- overall reduced demand for all forest products, affecting most significantly China, Japan, the Republic of Korea and Thailand;
- increased competitiveness of affected exporting countries (including Indonesia and Malaysia) through exchange rate depreciation, but in the face of reduced demand;
- falling prices for forest products throughout the region;
- reduced earnings in the forestry sector resulting in closure of mills, reduced harvests and workforce lay-offs.

These effects have spread outside Asia to a number of countries that have heavy trade dependence on Asian markets or are in competition with them.

The Asian crisis is of considerable importance to developing countries' trade in forest products. Most exports (by value) from developing countries are from Asian countries, which dominate developing country exports of wood-based panels, logs, sawnwood and paper products. South American countries, however, are the major developing country exporters of wood pulp.

Since Asia accounts for about 80 percent of tropical wood exports and more than 70 percent of tropical wood imports by value (Japan alone accounted for 42 percent of tropical wood imports in 1996), the Asian crisis is likely to disrupt trade in tropical wood products more severely than in non-tropical wood products. Weak demand in key markets and disruptions in key producer countries are likely to affect this trade most severely.

An important effect of the Asian difficulties has been changes to Indonesian forest policies resulting from the Indonesian Government's efforts to meet International Monetary Fund loan requirements. These include:

- reduction of export taxes on logs and rattan to a maximum of 10 percent by March 1998 (previously, very high export taxes made legal export of unprocessed logs or rattan almost impossible);
- elimination of the APKINDO (Indonesian Plywood Association) monopoly over plywood exports;
- establishment of new resource rent taxes on timber resources, including increased timber stumpage fees and a new auction system to allocate concessions.

In the short term, at least, adjustment to these conditions is likely to constrain Indonesian production.

The effects of the Asian crisis have been felt in trade in other markets and by exporters ranging from Africa to New Zealand to North and South America. Price drops and attempts by Asian exporters to maintain sales have resulted in falling prices for exporters in other (mainly tropical) regions. Market levels have dropped, but market shares have also changed as prices have fallen. African and South American exporters, for example, have been affected by lower Asian prices, the latter particularly by cheaper Asian plywood.

In summary, the main effects of the economic crisis have been lower demand, especially in Asian markets, disruption of supplies, weakened currencies in most developing export countries, lower prices and intense competition. Market prices for tropical products have fallen significantly. ITTO statistics indicate that prices of many tropical forest products fell by up to 60 percent in early 1998. The situation improved in mid-1998, but additional political and financial difficulties seen in the autumn of 1998 in many countries – including the

Russian Federation, China, many other Asian countries and some in South America – have caused widespread uncertainty in forest product markets.

Developments in trade liberalization

Moves to reduce trade restrictions that limit and distort free trade in forest products have continued to receive considerable attention.[30] Both developed and developing countries have begun to implement the global reductions in tariffs agreed to by the Uruguay Round in 1994. These tariff reductions have been supplemented by others stimulated by regional discussions such as the Asia-Pacific Economic Cooperation (APEC) and Southern Common Market (MERCOSUR) agreements. The terms of the Uruguay Round require that commitments to tariff reductions be fulfilled by 2004. Some countries have even advanced the dates for achieving these reductions. For example, Canada, Japan, the United States and the EU countries have agreed to eliminate all tariffs on pulp and paper products. As tariff rates on forest products in the main importing countries were generally quite low before the Uruguay Round, changes in these markets have not, in general, been substantial.

Some developing countries have also agreed to make substantial reductions in their tariff rates. Major reductions in forest products tariffs in India and negotiations on the accession of China to WTO are of particular significance. Removal of other trade restrictions, such as import licensing, has, for example, triggered the recommencement of the Indian log trade during the past three years. Similarly, negotiations with China on forest products tariffs appear to be close to settlement, and some relatively substantial reductions are likely; early reports suggest that Chinese tariffs on wood and pulp products may be halved (on average) while paper products may be reduced by slightly more (on average). The enormous populations and demand potential in India and China give these changes obvious importance.

[30] See FAO, 1998e for a detailed analysis of trade restrictions affecting forest products.

Improvements to non-tariff barriers are more difficult to identify, but the Uruguay Round agreements have given the subject more visibility. Several of the agreements subsidiary to the Uruguay Round settlement are important to the forest sector, most notably the Agreement on Technical Barriers to Trade and the Agreement on the Application of Sanitary and Phytosanitary Measures.

The most significant benefit related to the Uruguay Round agreements appears to be the impetus that the negotiations have given to continued reductions in tariffs and improvements in access.

At present, issues of general significance for the forest sector being discussed in WTO include those being considered in the Committee on Trade and Environment (see below) and the possible initiation of a new round of multilateral trade talks on trade liberalization (similar to the Uruguay Round). A number of countries are promoting a "Millennium Round" which would start in 2000. Others are of the opinion that the full implementation of the Uruguay Round agreements and the resolution of problems faced by developing countries should take place before any new negotiations begin. No agreement has been reached on this issue.

In addition to the reductions resulting from the Uruguay Round, an increasing number of regional trade agreements are establishing preferential terms of trade or, often, free-trade blocs. The list includes APEC, ASEAN, the Australia-New Zealand Closer Economic Relations Trade Agreement (ANZCERTA), the Caribbean Community and Common Market (CARICOM), EU, MERCOSUR, the North American Free Trade Agreement (NAFTA) and the South Pacific Regional Trade and Economic Cooperation Agreement (SPARTECA). All of these include reductions affecting the forest sector.

For example, APEC has proposed a timetable for the liberalization of trade across the region under which member countries are committed to creating a region of free and open trade and investment no later than 2010 for industrialized economies and 2020 for developing economies. In November 1997, the forest products sector was selected as one of the

15 sectors for Early Voluntary Sectoral Liberalization (EVSL), under which member countries will be asked to implement voluntarily an accelerated programme of trade liberalization. Development of proposals on this subject is expected to be completed by the end of 1998, with the main objective being to achieve a free market for forest products within APEC economies by the middle of the next decade.

As another example, preparations are currently under way to negotiate a Free Trade Area of the Americas, with negotiations planned to conclude in 2004.

Some concern has been expressed that a proliferation of such regional trade agreements may adversely affect global trade liberalization efforts. Another fear is that the effects of the Asian crisis and, in particular, the current problems in Japan will increase the difficulties faced by the trading system. A particular concern is that the economic downturn will result in a resurgence of protectionist measures as countries attempt to protect themselves against competition from lower-priced imports.

Trade and environment – impacts and developments

Although there is general agreement that trade and environment considerations should be mutually supportive, there are widely differing views on how far this can be achieved and how this objective might be approached. Other important issues of debate are what areas should be given most attention, and whether trade bodies or environmental bodies should predominate when conflicts arise.

The conjunction of trade and the environment was considered important enough that it was made one of the programme areas of the Intergovernmental Panel on Forests (IPF). The complex and contentious nature of the issues is illustrated by the fact that IPF could not reach agreement on many of them, and the subject was passed to the Intergovernmental Forum on Forests (IFF) for further consideration.

The lack of consensus continued when IFF met and again discussed these issues in September 1998. Subjects on which no agreement was reached include market access, trade and sustainable forest management, the relationship between obligations under international agreements and national trade measures, and illegal trade.

Efforts to clarify the issues and seek agreement will be made at a meeting on trade-related aspects of sustainable management of all types of forests, to be hosted by the Government of Brazil, UNCTAD and ITTO in February 1999. The topic will again be discussed by IFF at its third session in May 1999.

Certification (see the following section) is one of the controversial issues concerning trade and the environment. Producer countries and trade groups highlight its trade-restrictive aspects; consumer countries with strong environmental lobbies highlight the environmental advantages; and many consumer interests see eco-labelling[31] as a means of overcoming a variety of problems.

Some of the issues involved in the trade-environment debate are under further discussion in WTO's Committee on Trade and Environment (CTE); these include eco-labelling, market access, the effects of trade liberalization and the effects of WTO agreements (see e.g. WTO, 1998). Although CTE's main focus is on general trade issues rather than on specific sectors, forestry is emerging as one of the sectors where many of these issues come together, because most aspects of the forest sector have clear links to the environment. Some of the points being debated in CTE, most of which have relevance to forestry, are:

- the impact of trade liberalization on the environment;
- whether non-governmental actions such as certification, which are outside international trade rules, act as non-tariff barriers to trade;
- the closeness of the link between trade and the environment;

[31] Eco-labelling refers to a guarantee that certain environmental standards have been met at all stages leading to the final product, including production, transport and processing. It thus has broader coverage than certification, which indicates that the product comes from forests managed according to a defined environmental standard but implies nothing about postharvest practices.

- whether actions such as eco-labelling and certification affect trade positively or negatively;
- if actions that are taken for environmental reasons, and restrict trade, are contrary to WTO rules;
- which multilateral agreement should be dominant in the case of conflict – the multilateral environment agreement (MEA) or the multilateral trade agreement.

Forest products certification – an update

Certification of forest products continues to be a high-profile and often controversial subject in the forest sector. In addition, tentative efforts are being made to extend certification to non-wood forest products, which would raise new issues.

Accurate statistics on the area of forests and volume of wood certified are difficult to obtain, and the figures are often difficult to interpret. The Forest Stewardship Council (FSC) reports that about 10.3 million hectares have been certified by FSC-accredited certifiers.[32] It is significant that 90 percent of this area is in temperate developed countries, largely in Europe and North America. Sweden and Poland alone account for 58 percent of the total. Thus only a minor part is in tropical countries, where the problem of deforestation is greatest. The volume of wood involved and the volumes entering or about to enter the market are unknown, but are still insignificant in global and regional terms. The area of forests certified is not a sound indicator of the volume of wood entering the market, since parts of the certified areas may not be at harvestable age, may not contain commercial species or in extreme cases may not even have trees on them.

Certification efforts are being made at all levels. International efforts include those of FSC and the International Organization for Standardization (ISO); regional initiatives include those of the African Timber Organization (ATO) and EU's Eco-Management and Auditing Scheme (EMAS) and Eco-Labelling Scheme; and countries with national programmes include Brazil, Canada, Finland, Ghana, Indonesia and Sweden. In the last two years there have been many new initiatives, and a number of additional forests have been certified or are in the process of being certified. Among importing countries, interest continues to be greatest in Europe, especially Germany, the Netherlands and the United Kingdom. The exporting countries showing the most interest in certification are those whose main export markets are European countries and to a lesser extent the United States; hence the considerable effort Canada, Finland, Indonesia, Malaysia and Sweden have put into developing national certification systems.

Certification systems are based on evaluating the standard of forest management being practised. The two main approaches, whose relative merits are a subject of considerable disagreement, are those of FSC and ISO. FSC favours a performance-based approach, i.e. stipulating that a specified level of forest management, covering all aspects – including social aspects – must be achieved. ISO's approach is based on the management system, i.e. stipulating that specific management systems and commitment to specified actions and procedures must be in place. The FSC approach also considers chain-of-custody monitoring as an essential part of the process, while other certification approaches do not. Many consider chain-of-custody too difficult and expensive to contemplate.

Countries use one of these approaches or a modified version as appropriate to their own circumstances. While many consider the two approaches incompatible, there appears to be growing recognition that they may in fact be mutually supportive, and that a degree of mutual acceptance could eventually be achieved. It is also recognized that without this mutual acceptance, progress will be fragmented and difficult.

Even within an individual country, different approaches are being followed by different groups. Sweden, for example, recently announced that agreement had been reached between the large forestry companies and FSC on a certification system based on FSC's principles. Small forest

[32] As at 15 May 1998. In 1996, the area of forests certified, under both FSC and other certification schemes, was given as 2.3 to 4 million hectares (FAO, 1997d).

BOX 12
Certification and small forest owners

While not opposed to certification as a process, many small forest owners are opposed to the systems being proposed or to the way in which they have been developed. Many believe that the schemes have been designed to fit the circumstances of large forest owners and have not taken the special features of small-scale ownership into consideration.

The concerns of small forest owners include the following.

- Many principles and criteria of certification do not reflect forest management practices that are appropriate to small forests.
- Many social and environmental requirements are inappropriate to small owners (rather, they are relevant to large forest areas or the country as a whole).
- Tracking wood flows from small forests is difficult.
- Certification requirements discriminate against planted forests, especially small planted forests.
- Certification procedures and the control of the certification processes limit the freedom of small owners to make decisions about alternative land uses.
- Certification of small forests is prohibitively expensive; the cost for the initial certifying cost and the annual monitoring of small forests is substantially greater per hectare than for large forest areas.

These concerns have prompted efforts in a number of countries (e.g. Finland, France, Norway, Portugal, Spain and Sweden) to develop certification systems that are more applicable to small forest owners. In Finland's proposed national system, voluntary certification would be possible at three different levels: certification of individual forest owners; group certification in a Forest Management Associations area, which covers on average 80 000 ha and has some 1 500 owners; and group certification at the level of a Finnish Forestry Centre area, which comprises an average forest area of approximately 1.5 million hectares and has 30 000 owners. The registration authority in Finland would be the Finnish Environment Institute, while the accreditation body for the certification bodies would be the Finnish Accreditation Service.

owners in Sweden have rejected this system, however, and are following their own approach.

For many countries an important aspect of certification is its relevance and impact for small forest owners. This is an important issue in countries where a high proportion of the forest land is owned by a large number of owners, many of whom have very small areas of forest – often less than 10 ha. For example, in Finland 62 percent of the forest land is owned by some 440 000 private forest owners, with an average forest holding of about 26 ha. (Of the rest, the State owns 25 percent, large companies 8 percent and others, including municipalities, parishes, common forest owners and other bodies, 5 percent.) Similarly, in France, almost 10 million hectares of forests are owned by some 4 million private owners – an average holding of 2.8 ha per owner. Even in the United States, where large private companies own substantial forest areas, some 60 percent of the commercial forest land is owned by 10 million small-scale farmers and landowners.

Small owners are concerned that the certification systems being promoted may discriminate against them, restrict their freedom, exclude them from markets or be excessively costly to implement. They feel that many of the principles developed are either impossible for small producers to meet or inappropriate to their situation (see Box 12).

Many forest and paper processors use material from many small suppliers which is thus difficult or expensive to identify and track. It may also include waste, recycled or reused material. Materials from various sources are often combined in the manufacturing process, components are sold for remanufacture, and some products are recycled and manufactured into new products. The need to address these problems is resulting in continual modification of certification rules and procedures. Included in this adaptation is a move towards certification based on the proportion of certified material in the final product. This would allow certification of products without requiring that 100 percent of the material used comes from certified sources.

The following are some of the most significant developments that have occurred in the past two years.

- In Canada, an ISO-based certification system has been developed by the Canadian Standards Association (CSA). The system allows individual companies to be certified by a third party assessor using standards developed by CSA. No forests have been reported as certified under this system to date.
- In a related move, an ISO Forestry Working Group has prepared a technical report to assist forestry organizations in implementing the ISO-14001 Environmental Management System Standard. The report provides information on forest principles, criteria and indicators of sustainable forest management that have been developed by various other groups.
- The Netherlands and Germany have been working on the development of methods of validating the certificates provided by suppliers (both domestic and foreign) to their markets and of linking these certificates to final products in their markets. The Keurhout Foundation in the Netherlands has developed a system which will assess suppliers' certificates (from any certified source) and monitor the chain of custody (i.e. track the raw material through to the final product form) to the end user. A "hallmark stamp" (i.e. a label) is available for final products.
- The American Forest and Paper Association (AF&PA) has developed a Sustainable Forestry Initiative (SFI) by which its member companies commit themselves to move towards sustainable forestry. This commitment is a compulsory requirement of membership in AF&PA. It is not a certification process, but the principles and guidelines that companies espouse may serve as a basis for future certification by an independent body.
- FSC has continued to expand its country coverage and has developed an FSC label to be placed on certified final products. This label is available for use on supplies certified by FSC-accredited certifiers.

Despite these efforts, there are still unresolved or unanswered questions surrounding certification and uncertainty as to how it may develop in the future. There is still little evidence of the market impacts of certification, positive or negative. The most obvious sign of growth in certification activities is the area of forests certified, but there are few signs of significant volumes of certified products entering the market. In part, this is because certified supplies remain limited, but it may also reflect a lack of buyer interest.

It remains unclear whether demand for certified wood will increase, and whether a price premium is likely. Even in the markets showing the greatest interest in certification, there is little sign of a substantial or increasing demand, or any price premium paid. A number of market studies (e.g. Brockmann, Hemmelskamp and Hohmeyer, 1996) have concluded that there is little evidence of a significant demand for certified products.

Another unanswered question is whether certification will, in fact, significantly contribute to improved forest management where deforestation is greatest – in the developing countries. Certification was originally promoted by environmental groups as a market-based lever to improve forest management and reduce deforestation. At present, those certifying or trading in certified products seem to be using it more as a marketing tool either to generate an increasing market share or to ensure continued or improved access to markets. Some forest owners, however, see certification as benefiting their forest management practices in addition to providing market benefits (see Box 13).

Despite these uncertainties, interest in certification continues to grow. It seems clear that at least in the short to medium term the area of forests being certified will continue to expand as major producing countries such as Canada, Finland, Indonesia and Malaysia finalize the systems they are developing. It is still difficult, however, to predict where certification will finally arrive and what type of system, or combination of systems, will be favoured. The results will depend heavily

BOX 13

Why Precious Woods decided to apply for FSC certification

In 1997, the Brazilian subsidiary of Precious Woods, Madeireira Itacoatiara Ltda. (MIL), was certified in accordance with the principles of the Forest Stewardship Council (FSC). The decision to go ahead with certification was based on the company's expectation that although certification meant a long-term (at least five years) commitment of staff and finances, it would provide the company with both internal improvements and market benefits. Precious Woods felt that certification was a means of highlighting the company's long-term commitment to ecological and socially correct behaviour and could potentially bring the following benefits.

A marketing advantage

With the trend in some of the company's target markets towards boycotts or restrictions on tropical timber, the company felt that it was important to have a "green label" which indicated its use of ecologically sound harvesting methods. The certification process was seen as likely to facilitate the company's marketing activities. A revival of markets for tropical timber products was considered possible, both in completely new markets and in markets that were lost during boycott campaigns. A certificate would assist the company in exporting tropical timber to ecologically sensitive markets, particularly those in central and northern Europe. Certification was seen as a means of indicating to the public the company's commitment to sound management of its tropical forests. It was felt that making certification methods, results and conditions regularly available to the public would provide promotional benefits.

Internal company improvement

The company foresaw internal benefits from the assessment process and from having its major forest activities, especially those in ecologically sensitive forest areas, checked by outside independent experts. The evaluation by external assessors was expected to provide a good opportunity for the company to modify its operations in order to minimize ecological problems. It was felt that the assessment process, carried out over a significant period, would benefit both the company and the certifier.

Source: Text provided by Precious Woods (Switzerland) Ltd.

on which markets institute it; the degree of support given by consumers, legislators and traders in these markets; and the extent to which harmonization, or at least mutual recognition of different practices, is achieved. Certification may expand and have a significant impact in some markets; equally, it could remain limited to a few markets and a few specific end uses (e.g. high-value furniture). It is also possible that in the long term it could fail to have any significant impact. The deciding factor will be consumer reaction to the products, which is far from clear at this stage.

FUTURE SUPPLY AND DEMAND FOR INDUSTRIAL ROUNDWOOD AND WOOD PRODUCTS

FAO has recently completed two new global supply and demand studies: the Global Fibre Supply Model (GFSM) (FAO, 1998a) and the Global Forest Products Outlook Study (FAO, 1999). A major and more detailed forestry sector study for the Asia and the Pacific region has also recently been produced (FAO, 1998f). Rather than dwelling too intently on market forecasts which will be vulnerable to macroeconomic shifts (such as those recently experienced in Asia), these new studies focus more on the likely policy implications of forest product market developments. The following text briefly describes the main market developments that are expected in the future and discusses the implications of these developments within the context of sustainable forest management.

Future forest product production and consumption by region

The current and projected level of wood production and consumption in 2010 is shown by region in

FIGURE 4

Forecast industrial roundwood production and product consumption (converted to equivalent roundwood input) in 1996 and 2010

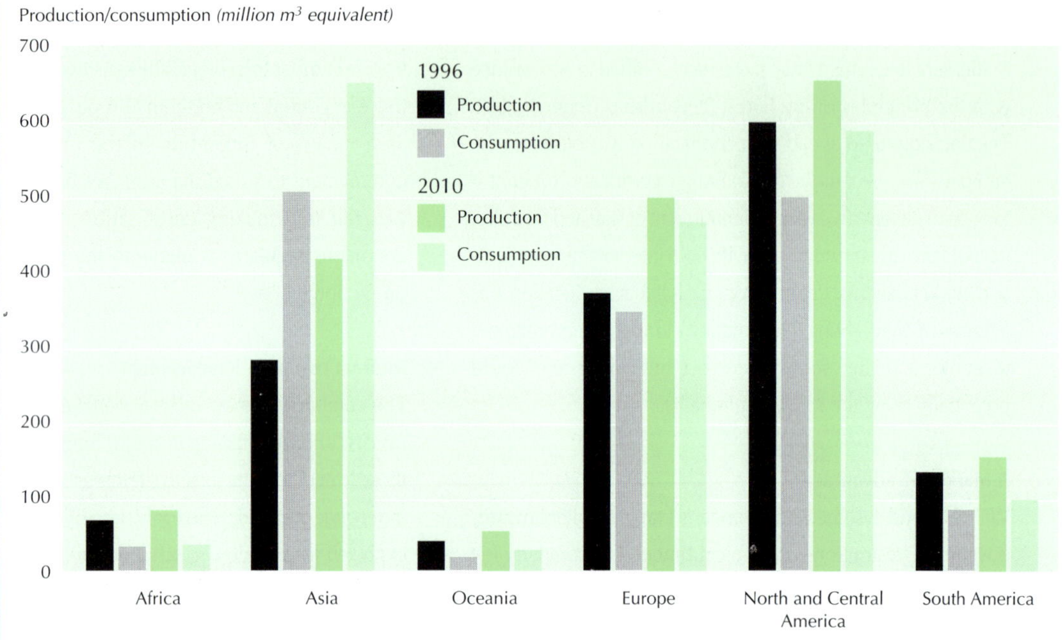

Figure 4. (Current levels of wood consumption and production by product and country are given in Table 4 of Annex 3.) From 1996 to 2010, global industrial forest product production and consumption are projected to increase at an annual rate of about 1.7 percent, from 1 490 million to 1 870 million cubic metres. Thus in 2010, output will be about one-quarter higher than it is at present. It will, however, only be about 10 percent higher than the peak in production (1 700 million cubic metres) around 1990.

Growth will vary between regions, with Asia and Oceania likely to show the highest rates of expansion. Obviously, the actual developments that transpire in Asia will supersede any attempts to model the outlook for forest product markets (see Box 14). Slow growth in consumption is expected for Africa and South America, and slow growth in both consumption and production is expected for North and Central America. North and Central America will, however, remain by far the largest producing and exporting region in the world.

Europe, Asia and North and Central America will account for about 85 percent of production and over 90 percent of consumption in 2010 (roughly the same share as in 1996). However, within this group it is expected that a small share (about 5 percent) of global consumption will be gained over the period by Asia at the expense of North and Central America.

In terms of net trade (the difference between the heights of the consumption and production bars in Figure 4), Asia will continue to be the world's only net roundwood and product importing region. In terms of gross trade flows, however, trading patterns are not expected to change, but the shape and form of trade will continue to change. Since the 1950s, trade in forest products has steadily increased as a proportion of total production. The outlook is for more trade, both in gross volumes and as a proportion of production.

Increases in trade are predicted for two reasons. First, many countries are expected to continue to give priority to developing manufacturing and processing capability rather than exporting roundwood and pulp. This will prompt continued declines in exports of semi-processed products as producers seek to add value to the raw material.

BOX 14
The impact of the recent Asian crisis on supply and demand projections

The model used to produce the supply and demand forecasts of the Global Forest Products Outlook Study was also used to estimate the potential long-term impact of the recent Asian crisis on global markets. The estimate looks at the consequences for the forestry sector up to 2010. No attempt was made to forecast near-term market fluctuations.

In April 1998, analysts were projecting a resumption of economic growth in Asia after a period of retrenchment whose length and severity were in question. Under a projection of more measured economic growth in Indonesia, Malaysia, the Republic of Korea and Thailand, plus knock-on effects in some other countries in the region, overall consumption in the region across all wood product categories is estimated to be roughly 4 to 5 percent lower than the baseline projection for the year 2010.

Slower economic growth in these countries, compounded by the effects of competitive devaluation, is expected to suppress imports as a share of overall consumption in the region, which has been such a large importer. Countries in the region that are less affected economically by the crisis are likely to increase forest product imports if offshore prices fall, although this could come at the expense of domestic production. These counterbalancing effects illustrate the difficulties of modelling structural adjustment.

In general, the outlook remains murky because of the difficulty of foreseeing the types of currency adjustments that may emerge and their competitive effects on trade. Compounding the confusion in this type of assessment is the inability to understand or project the impacts of the financial crisis on the availability of capital to the region. Any outlook for growth is founded on the assumption of ample supplies of development capital at market clearing prices. The trademark of this financial crisis is a dramatic shift and reassessment of the role of global capital markets and the respective roles, responsibilities and liabilities of governments, financial institutions, entrepreneurs and investors. Until these roles are clarified, the supplies of financial capital to the region may remain constrained. Under a scenario of scarce capital, forest products output in the region is likely to show little growth, and consumption only very slight increases.

For example, a greater proportion of commodities such as sawnwood and panels will be processed further into furniture and joinery products. Second, an expansion of domestic markets is expected as developing countries' economies grow and mature. Maturation of these markets will give rise to economies of scale in processing, product design, assembly, manufacture and distribution. The increasing specialization, market segmentation and competition will promote higher levels of trade, both within regions and internationally.

On the supply side, the factors leading to greater production of both roundwood and products in the future will differ widely among countries. In some countries, roundwood production will be increased by supply-side factors, such as the opening-up of new areas of forest and the maturation of extensive plantation resources (e.g. in countries around the southern Pacific Rim). However, more typically, production will be driven up by increased demand as a result of high rates of economic growth (e.g. in most European and many Asian economies). In countries where there is considerable demand pull, and particularly where forest resources are limited, wood product producers will start to consider using a broader range of wood and fibre raw materials than they have in the past (see Box 15).

Future forest product production and consumption by product category

Current estimates and future forecasts of global forest product production and consumption by product category are shown in Table 7. As in the past, the market for paper and paperboard is expected to have the most rapid growth, at an annual rate of 2.4 percent (see Table 7 and Figures 5 to 10). In contrast, production of pulp for paper is expected to grow by only 1.1 percent per year, reflecting an expected increase in the use of

BOX 15

The wide range of potential wood and fibre resources in the Asia and the Pacific region

An analysis of the supply outlook for all potential sources of wood and fibre supply was carried out as part of the Asia-Pacific Forestry Sector Outlook Study. This analysis incorporated information from several studies on supply potential from natural forests, plantations and trees outside forests and on harvesting and wood processing residues, recycled fibre and non-wood fibre produced in the region each year. The main findings are shown in the figure.

The forecasts were made on the basis of existing technology (except for a trend towards more recovery of wastepaper in the future) and policies (e.g. with respect to the area of forest in legally protected areas). Historical trends in forest conversion to other land uses were also incorporated in the forecast. The horizontal lines in the figure represent projected production of recycled and non-wood fibre, pulpwood, sawlogs and fuelwood in the region in 2010.

As the graph shows, the region has a large potential to produce sawlogs and other fibre from outside the areas that would be typically considered in wood supply and demand analysis (i.e. natural forest and plantations). In terms of other fibre production potential, non-forest sources far exceed the potential of the forest to meet production needs. For example, trees outside the forest have twice the potential of forest plantations to produce small roundwood, because of the large area of agricultural land (particularly agricultural tree crops) in the area. However, few reliable statistics on trees outside forests are available, so the exact magnitude of this resource is uncertain. Recovered paper and wood processing residues could also meet the region's entire needs for pulpwood. In terms of sawlog production, industry has to look to forests for high-quality logs; about half the potential sawlog production in the region is from the natural forest and a further quarter from forest plantations. However, trees outside forests account for the remaining one-quarter of sawlog production potential and even in this category could go a long way towards meeting production requirements.

Several countries with limited forest resources (e.g. many countries in South Asia) already use a wide variety of sources for sawlog and fibre supply. Countries that currently rely on natural forests for much of their needs may choose to continue to do so.

Sources: FAO, 1997j; 1998a; 1998f.

TABLE 7

Current and forecast global forest production/consumption by product category, 1996 and 2010

Product	Production/consumption 1996	Production/consumption 2010	Total growth 1996-2010 (%)	Annual growth 1996-2010 (%)
Industrial roundwood *(million m³)*	1 490	1 872	26	1.7
Sawnwood *(million m³)*	430	501	17	1.1
Wood-based panels *(million m³)*	149	180	20	1.3
Pulp *(million tonnes)*	179	208	16	1.1
Paper and paperboard *(million tonnes)*	284	394	39	2.4

Potential production of sawlogs and other fibre from different sources in the Asia and the Pacific region in 2010

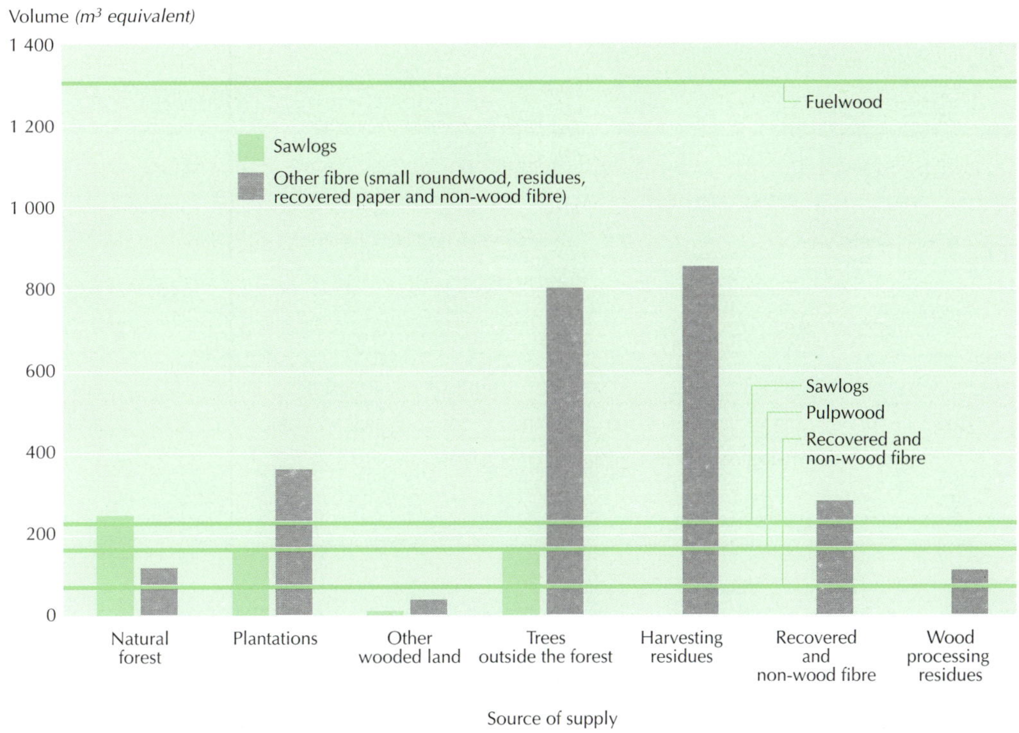

recovered paper in the total fibre furnish in the future. Moderate growth is expected in solid wood product consumption, at annual rates of 1.1 percent for sawnwood and 1.3 percent for wood-based panels. Most of the growth in the production and consumption of wood-based panels is expected in the reconstituted wood panels sector rather than the plywood sector.

Forecast wood and fibre production compared with production potential

The comparison of forecast roundwood production levels in the Asia and the Pacific region with production (or estimated biological) potential (Box 15) shows that, in general, future wood demand could easily be met within the region. In certain countries and for certain types of wood (e.g. sawlogs), however, supplies are going to become increasingly scarce in the future.

FAO does not yet have sufficient data to make an accurate assessment of the production potential for the whole world. The Global Fibre Supply Model, for example, covers a large part of the world but excludes the important contribution of trees outside forests. However, GFSM can be used to compare projected levels of production (as shown in Figure 4) with supply potential from the forest and recovered and non-wood fibre sources across some regions (Table 8).

As Table 8 shows, forecast production levels are well within the forecast limits of production potential in South America and Oceania, but approach the limit in Asia and exceed it in Africa. The results of the fifth European Timber Trends

FIGURE 5
Current and forecast supply and demand – North and Central America

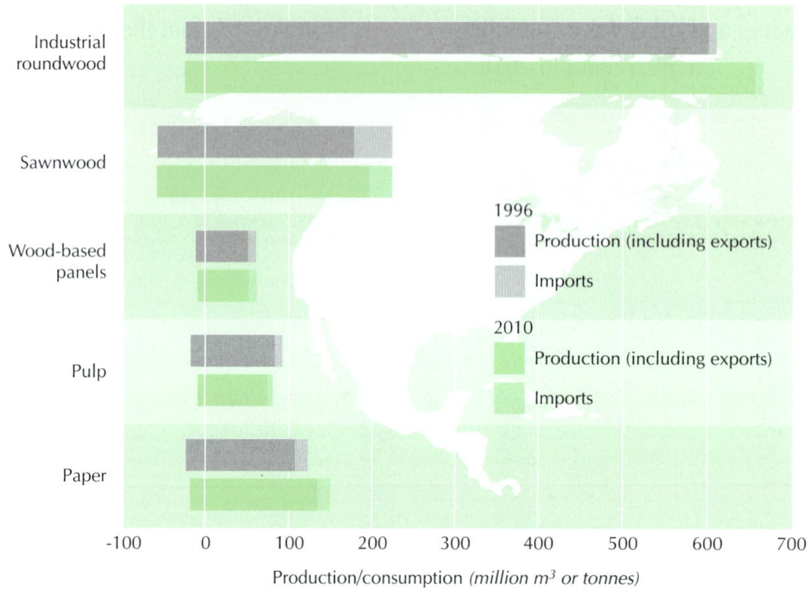

Note: Negative figures represent the volume of exports.

FIGURE 6
Current and forecast supply and demand – South America

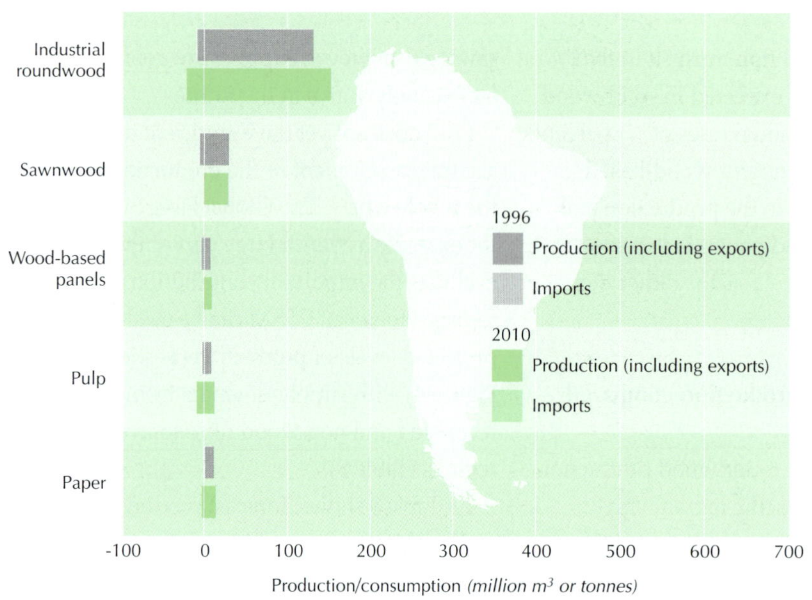

Note: Negative figures represent the volume of exports.

FIGURE 7
Current and forecast supply and demand – Europe

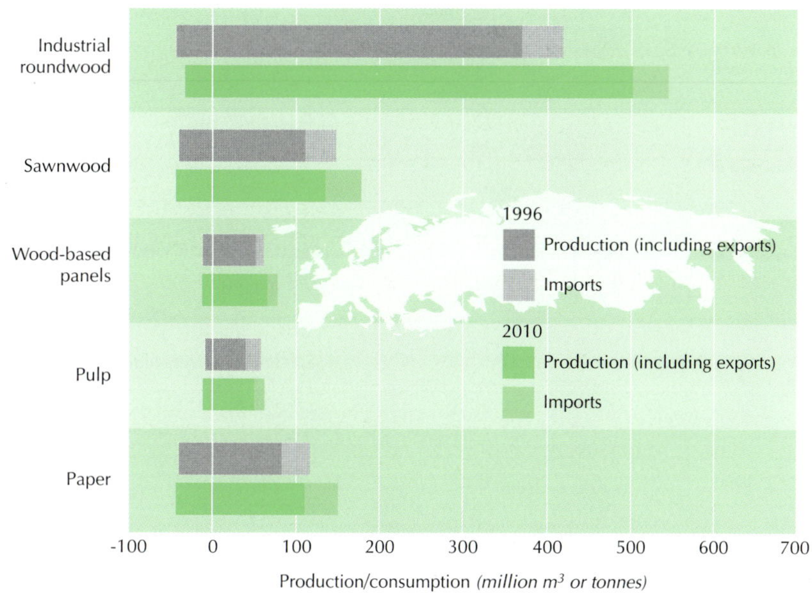

Note: Negative figures represent the volume of exports.

FIGURE 8
Current and forecast supply and demand – Africa

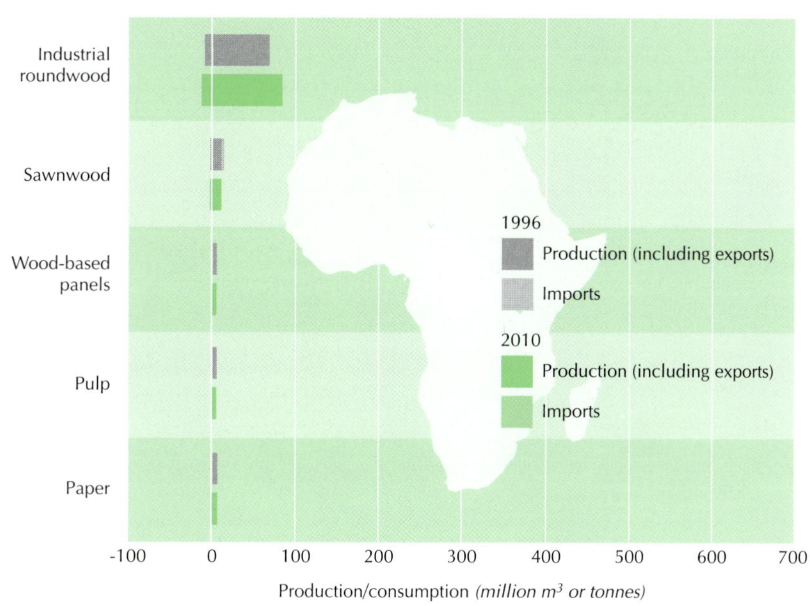

Note: Negative figures represent the volume of exports.

FIGURE 9
Current and forecast supply and demand – Asia

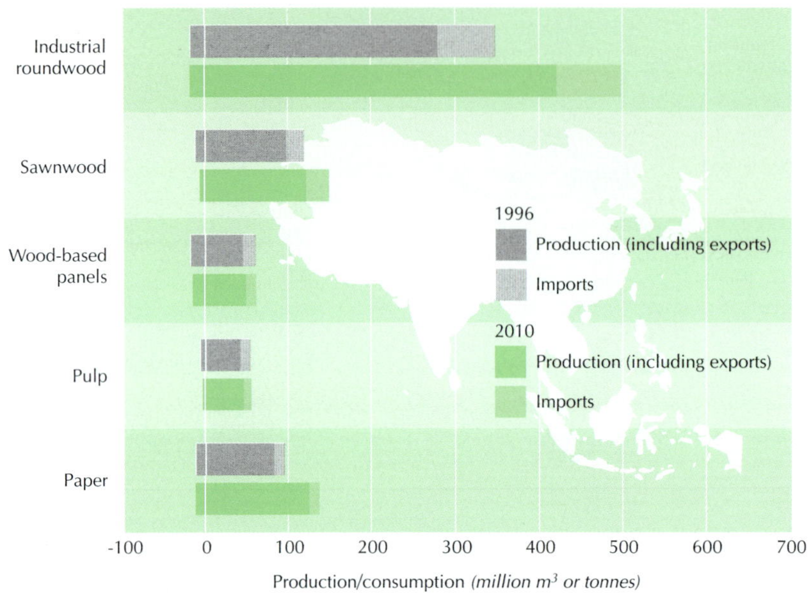

Note: Negative figures represent the volume of exports.

FIGURE 10
Current and forecast supply and demand – Oceania

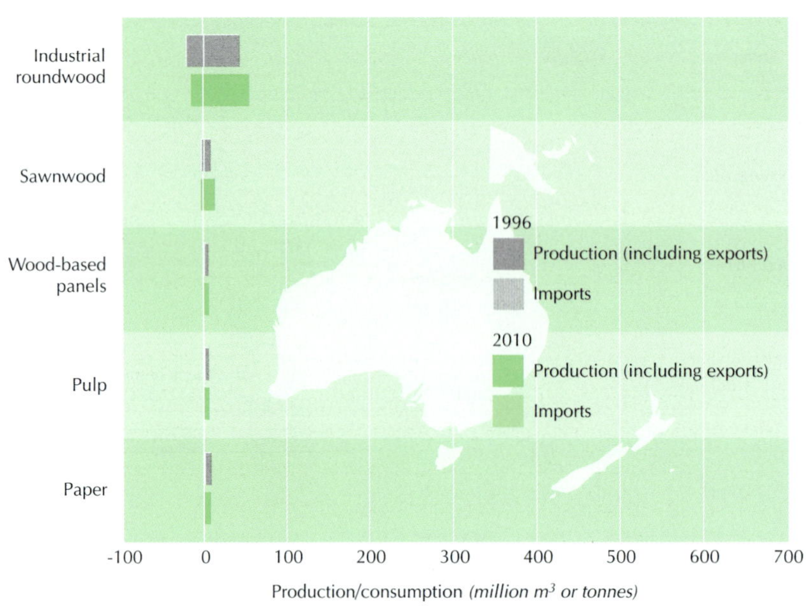

Note: Negative figures represent the volume of exports.

TABLE 8

Forecast production potential from forests and recovered and non-wood fibre sources and forecast production of wood and fibre in 2010 (million m[3] equivalent)

Region	Forecast production in 2010			Total potential fibre availability in 2010[a]
	Industrial roundwood	Recovered and non-wood fibre	Total	
Africa	84	2	86	81
Asia	421	222	643	729
Oceania	54	0	54	80
Europe	502	133	632	893
North and Central America	658	147	805	835
South America	153	2	155	225
World total	**1 872**	**506**	**2 375**	**2 843**

[a] From GFSM (FAO, 1998a) or official government estimates.

Study (Joint ECA/FAO Agriculture and Timber Division, 1996) suggest that Europe should also have adequate wood supplies to meet production requirements in the near future. However, two points are worth noting. First, the GFSM supply results should be considered as the absolute maximum amounts. The cost of accessing increasingly marginal areas which are included in the GFSM analysis may prevent the total potential supply presented in Table 8 from being utilized in the near future. Second, it should be remembered that, while supplies may be plentiful at the broad regional or country level, there may continue to be local scarcity which could put forestry policy-makers under pressure to release areas of natural forest for timber harvesting.

In regions such as Africa and Asia, where supplies (particularly of large logs) from forests are coming under pressure, consumers of wood and fibre will increasingly have to look to other sources to meet the demand (as they already do in Asia) if they do not wish to overexploit the forest. Another, perhaps more likely, alternative is that the markets for forest products will continue to move in the direction of substitution of sawnwood and plywood by other wood-based panels and engineered wood products, which can be manufactured from small-sized wood or non-wood substitutes.

To conclude, the supply and demand analysis carried out by FAO suggests that the capacity of the forest and other sources of fibre will be sufficient to meet demand for the foreseeable future. However, the situation will vary among countries and regions. Africa and South Asia will continue to have to use a wide range of non-forest supply sources to meet their needs. The demand for higher-quality sawlogs will also approach or even exceed the production capacity of forests and plantations in Africa, Southeast Asia and the Pacific Islands.

It is expected that product prices will not rise significantly over the projection period. Many regions have ample or excess wood product manufacturing capacity. The current worldwide economic slowdown will further mitigate pressures on consumer prices. In selected cases, there may be upward pressure on the prices of certain types of roundwood (typically the higher grades), but price increases will be restrained by the availability of cheaper wood and non-wood substitutes. Trading patterns are not expected to change significantly, apart from the continuing trend towards more in-country processing of wood raw materials. This should lead to less trade in semi-processed and commodity-grade wood products and more trade in higher-value products.

Possible future market developments

Wood supply. The sources of wood and other fibres used in production are likely to change in the future. It is expected that in most countries there will be a move away from the use of forest resources for wood and fibre production towards other land-based and non-land-based sources of supply. The greatest change by far will be the increased use of wood processing residues and recycled fibres in the product input mix. The use of such secondary sources is likely to continue to expand in the more developed parts of North America, Europe and Asia, while trees outside forests are likely to have an increasingly important role as forest resources decline in some of the less developed regions of the world.

Within the forest, supply patterns are also likely to change in the future. The next ten years or so will see large areas of commercial short-rotation plantations (for pulpwood) come on stream in the Southern Hemisphere (see discussion on plantations above). Greater areas of older plantations established for the production of sawlogs will also start to be harvested in countries such as Australia, Chile, New Zealand, South Africa, the United Kingdom and the United States. These plantations will provide the greatest share of the expanded wood production potential expected in the future. In contrast, very few countries are likely to be able to expand production sustainably from the natural forest without considerable investment in silviculture.

Greater areas of natural forest are likely to be made legally protected areas. Many of the areas that are likely to be chosen for preservation are not currently harvested and are considered unexploitable for economic reasons (physical constraints, transport limitations, low timber value). A decrease in harvesting intensities in the exploitable natural forest would have a greater impact on future supply potential. Harvesting intensities might fall in the future for two reasons:

- tropical countries may modify existing harvesting regulations to reduce the volume of timber that may be felled in a given area, in response to environmental concerns;
- stocking levels may fall as forest operations move out of virgin forest and into secondary forest.

The combination of these factors could have a dramatic effect on future timber availability and will reinforce the expected switch from natural forest to plantations and non-forest supply sources outlined above.

Technological change. Technological change has been incorporated in the above supply and demand analysis only in the pulp and paper sector, where it has been assumed that current trends in the use of recovered paper in the total fibre furnish will continue in the future. For example, 1 tonne of paper and paperboard in 1970 was made up of over 80 percent wood pulp. By 1997, this figure had dropped to 56 percent and by 2010 it is expected to fall to below 50 percent. This trend is partly a result of the increased use of recovered paper, but is also a result of shifts in market shares, whereby the proportion of printing and writing papers (which have a lower fibre content) in the overall market has increased and is expected to continue to do so.

Other technological changes which the model has not included may also occur in the future. First, improvements in harvesting practices could increase log recovery and reduce logging residues in many of the world's forests. Many developing countries have substantial scope to increase their log recovery rates. Even a modest increase in log recovery rates in countries with high annual felling levels could increase production and contribute significantly to meeting the projected growth in industrial roundwood demand.

Better mill recovery rates could also have significant effects by reducing the amount of roundwood required to manufacture products (see Box 16). In addition, residues could be used more effectively to meet the demands of other wood processors. Not much is currently known about the

BOX 16

Recent developments in forest products processing with potential to decrease use of raw materials

Technological developments in wood processing continue to keep pace with market, environmental and raw material resource trends. The increased production of reconstituted panels such as oriented strand board and medium-density fibreboard and developments in the manufacturing of engineered wood products will continue to increase the efficiency of using raw material. Engineered wood products increase opportunities for using small-diameter logs of lower quality and less-used species. The use of microprocessors in almost every step of the production cycle increases product quality and minimizes the amount of residues generated.

Much of the technological development in wood processing is concentrated in industrialized countries, and the extent to which the technologies are used in non-industrialized countries is not known. Wider adoption of current technologies and development of new ones which increase processing efficiency could help slow the increasing demand for raw materials to supply finished wood-based products. Below are some examples of technologies that illustrate the opportunities.

- Technology developments in laminated veneer lumber include increasing use of ultrasonic veneer graders so that a substantial part of the veneer supply is routinely tested for possible use as laminated veneer lumber, rather than only for plywood. This development has contributed to a 75 percent increase in production of laminated veneer lumber and a threefold increase in I-joists in the United States from 1990 to 1996.
- New oriented strand board technologies include more efficient glue distribution on the strands, which makes it possible to achieve target properties with lower mat densification than before; larger panel sizes, resulting in less trim waste; new dryers that break and pulverize strands less; and widespread adoption of long log flakers which reduce log trim and kerf loss. Mills using the older processes achieve recoveries estimated at around 55 to 60 percent, while those using the newer ones report recoveries of 60 to 63 percent (ratio of wood volume in to product volume out).
- New technologies developed in medium-density fibreboard (used particularly for furniture components) include three-dimensional thermoform laminating, surface printing and wet finishing techniques. The production of medium-density fibreboard in Europe increased threefold from 1990 to 1997.
- The increasing volume of small-diameter material at sawmills is driving some technological developments. Mills specializing in manufacturing narrow-dimension lumber from small-diameter stems are adopting curve sawing. Scanning and optimization at the primary headrig and the secondary breakdown centres (edging, trimming, resawing, cant breakdown) are becoming standard. The manufacture of finger-jointed studs from small lengths salvaged from jacket boards and slabs which would have been chipped in the past is becoming increasingly common.
- The use of fibre other than wood for the production of particleboard is increasing. Plants using wheat straw for production of boards have been built in the United States and Canada within the past decade. Particleboard plants using cotton stalks have been built in India. Medium-density fibreboard plants are operating in Malaysia using oil-palm residues as a raw material. The product is based entirely on fibre strands of the oil-palm's fruit husks, which are a residue from the processing of the fruits. The characteristics of the fibre strands are comparable to those of local timber and the strands can be converted to quality panel products.
- The major development in the pulp and paper sector is a continued increase in the use of recovered paper. Increased customer acceptance of recycled paper and more efficient de-inking systems have contributed to the increased proportion of waste paper being reused. Other technologies which may result in decreased fibre demand per unit output are carbonation in reactor tanks, which improves the utilization of non-wood fibres by allowing the recovery of cooking chemicals (previously quite difficult because of the silica in the plants), and increased use of high-yield mechanical pulps, which have higher yield than chemical pulps.

utilization of mill residues outside a few of the large developed countries. However, it is suspected that large volumes of residues are wasted or left unused. As the Asia-Pacific Forestry Sector Outlook Study has shown, all of these sources could make a significant contribution to wood supply.

A third technological change which might occur is a move towards greater use of reconstituted panels because of two factors: upward price pressure resulting from the increasing scarcity of large-diameter logs, and technological developments in construction and other wood-using industries which would allow the use of such products where only plywood or sawnwood are currently used. This shift will also have the effect of extending the use of resources (recovery rates for reconstituted panels are typically higher than for sawnwood and plywood) as well as providing a ready market for residues from other industries.

Policy implications

The preceding analysis has discussed two broad structural changes expected to take place in the forest sector in the future. It is not possible to be precise about when and where these changes will take place (many of them already have occurred in Europe, for example), but it is certain that they will occur at some time. The important question for many forest policy-makers around the world is how to manage the change process. The analysis of future timber markets would suggest that three topics – pricing, human resource development and structural change – deserve the immediate attention of policy-makers. (See Part II for discussion of recent policy and institutional changes in the sector.)

Pricing. Prices are a powerful indicator of scarcity, and when they are wrong they can lead to serious misallocations of investment and resources. About 40 percent of the world's timber supply comes from private forests, and probably about the same proportion of supply is sold in competitive markets (or in a way that approximates competitive markets). However, governments largely control the pricing of wood extracted from the remainder of the forest (mostly natural forest), and prices are often set low to stimulate industrial development.

Setting low stumpage rates may satisfy certain development objectives but also often leads to undesirable effects, some of which are beginning to appear in many countries now. Underpricing of the resource discourages efficiency in harvesting and processing, reduces the incentive to invest in plantations and places alternative suppliers such as smallholders and recyclers at a disadvantage. Thus, designing better pricing policies will be crucial for stimulating the broadening of supply sources and for encouraging the efficiency improvements that will be required in the future.

One of the greatest challenges that forest policy-makers and forest managers will face in the future will be to generate the revenues necessary to finance sustainable forest management. Competitive market-based pricing of the resource will be an important first step in this process. Forestry policy-makers should consider how they can create competitive markets for the roundwood extracted from the natural forest so that the levies they set reflect the market value of the resource and remove the distortion in favour of harvesting wood from the natural forest.

Human resource development. Forestry is a labour-intensive activity. In order to introduce technological improvements and to meet the increasing demands placed on the sector for better standards of management and harvesting, the level of skills in the sector will have to be increased dramatically.

This is a massive task considering how many people are employed in the sector. For example, employment in industrial forestry has been estimated at approximately 1 million full-time equivalent (FTE) jobs in developed countries and 2.7 million FTE jobs in developing countries (Poschen, 1997). Furthermore, the need for upgraded skills does not stop at the production sector. An increased range of supply sources and desirable improvements in management, such as community involvement, will also generate a need for significant upgrading of the skills of staff in many countries' forest administrations and other organizations involved in forest management.

Structural change. The final pressing need of the sector will be an increased capacity to cope with the structural changes that are likely to take place. Many governments already have a wide range of measures designed to stimulate the development of certain types of domestic wood processing industries (e.g. preferential tax breaks, conditions attached to concession contracts, export bans). These should be reviewed to take into consideration the expected changes in patterns of wood supply (fewer large logs from the natural forest, more plantation-grown wood, more recycled paper). In addition, governments should consider the levels of investment that will be needed to finance the structural changes.

As noted above, forest administrations are going to be expected to deal with a wider range of issues in the future. To increase the social and environmental performance of the sector, sections within forest agencies that deal with conservation, community relations, watershed management and extension will be likely to require more resources in the future. Changes are likely to be required within the traditionally powerful harvesting and utilization sections of forest administrations. Officials concerned with wood supply will have to adjust to a move away from having strong control over a small number of concessionaires, towards having much less control over a larger number of smaller suppliers. They should consider devolving some aspects of control to regional, local and even community-based authorities, to cope with these expected changes. ◆

PART II
POLICY, PLANNING AND INSTITUTIONAL ARRANGEMENTS

Issues in national forest planning, policy and legislation

DEVELOPMENTS RELATED TO NATIONAL FOREST PLANNING

Progress in national forest programmes

The significance of comprehensive national forest sector planning processes and an enabling policy environment to sustainable forest management is now widely recognized. Various sector planning efforts around the world are focusing to a significant extent on a common set of principles and approaches, characterized as national forest programmes (NFPs).

"National forest programme" is a generic term for a wide range of approaches used by countries in planning, programming and implementing forest activities. Among the strategic frameworks considered as NFPs are: national forestry action plans started under the Tropical Forests Action Programme or the later country-led national forestry action programmes; forestry sector master plans; forestry sector reviews; national plans to combat desertification; national biodiversity strategies; national environmental action plans; national environmental management strategies; national conservation strategies; and forestry or forest components of national sustainable development strategies or national Agenda 21 strategies. Some 128 countries have developed or updated their NFPs within the past 13 years, and 13 countries have initiated NFPs within the past two years (see Table 5 of Annex 3).

These strategic frameworks represent means of ensuring consistency between national forest plans and programmes and the recommendations of the United Nations Conference on Environment and Development (UNCED) concerning sustainable forest management. NFPs have been endorsed by the Intergovernmental Panel on Forests (IPF) and the Intergovernmental Forum on Forests (IFF).

IPF endorsed the basic principles of NFPs as identified by FAO (see Box 17). It also emphasized the importance of considering certain factors in implementing NFPs, as follows:

- use of appropriate participatory mechanisms to involve all interested parties;
- decentralization where applicable and empowerment of regional and local government structures;
- recognition and respect for traditional and customary rights of, *inter alia*, indigenous groups, local communities, forest dwellers and forest owners;
- secure land tenure arrangements;
- establishment of effective coordination mechanisms and conflict-resolution schemes.

The basic principles of NFPs as defined by FAO (1996c) were also endorsed by IPF.

The second session of IFF recognized NFPs "to be a viable framework for addressing forest sector issues, including implementation of the IPF's proposals for action in a holistic, comprehensive and multisectoral manner". IFF agreed on the importance of mobilizing new and additional resources to support NFPs in developing countries and in countries with economies in transition, in

BOX 17
Basic principles of national forest programmes

- National sovereignty and country leadership
- Consistency with national policies and international commitments
- Integration with the country's sustainable development strategies
- Partnership and participation
- Holistic and intersectoral approaches
- Policy and institutional reform and capacity building

particular countries with low forest cover and with fragile forest ecosystems. IFF has asked FAO to present the results of a survey on the status of implementation of NFPs in all countries concerned to the third session of IFF in May 1999.

Improved accessibility of information on forests

The importance of accurate data and information on forest resources for national forest planning and global policy dialogue is increasingly recognized. Intensified efforts are being made to identify information gaps, to improve the information base, to harmonize information from different sources and to strengthen national capacities in information collection and analysis. Information management and dissemination are also regarded as increasingly critical. A current initiative to address the problem of information availability is the Global Forest Information System (see Box 18).

As important as reliable information is to efforts in sustainable forest management, lack of information should not be used as an excuse for inaction. Perfect information will never be available. The collection and management of information is often expensive. Every country must make choices about what information to collect and must weigh the potential benefit of the information with the cost of collecting and managing it. Some countries will simply be unable to afford the information they would ideally like to have. Particularly for these countries, it will be important to develop techniques for planning and management in "information-poor" environments.

Even as the information base improves, it is important to recognize that information is only a tool for analysis. It is the analysis that is critical for sector planning and policy formulation. The capacity of countries to manage and analyse information and to deal with varying interpretations of the same data will become increasingly important as information becomes more accessible.

BOX 18
The Global Forest Information System

The final report of the Intergovernmental Panel on Forests (IPF) underscored the need to improve access to information on forests: "The Panel emphasized the need to review and improve information systems. Attention should be given to worldwide access to information systems that would encourage effective implementation of national forest programmes, increased private-sector investment, efficient development and transfer of appropriate technologies, and improved cooperation" (Commission on Sustainable Development, 1997).

In 1997, several organizations that gather and publish information on forests began to examine the possibility of collaborating on a global forest information system. The proposal for the system responded to the need for accurate, current and accessible information on forest resources. It implicitly recognized the growing significance of electronic information systems, specifically the Internet, as a logical vehicle through which this information could be shared.

To advance this idea, the Center for International Forestry Research (CIFOR) and the World Conservation Monitoring Centre (WCMC) hosted a Workshop on Provision of Forest Information on the Web at the eleventh World Forestry Congress in October 1997. The concept of a global forest information system was well received and the organization of a task force to develop it was recommended. Subsequently, the Executive Board of the International Union of Forestry Research Organizations (IUFRO) directed one of its working groups to organize a task force. The task force prepared a proposal for the establishment of the Global Forest Information System. The proposal was discussed and supported by the International Consultation on Research and Information Systems, which was held in Ort-Gmunden, Austria in September 1998 as an intersessional meeting of IFF. In looking at the range of users and their extensive demands, one thing is quite clear: as society gains a better understanding of the value and role of forests, the numbers of those seeking forest information will grow, as will the list of their requirements.

RECENT POLICY ISSUES

Rapidly evolving international economic relationships, such as trade liberalization and the dynamic movement of capital and corporations across borders, and above all, national policy concerns, continue to drive national policy agendas.

Despite great variability among countries, three broad shifts are apparent. National policy-makers have become more aware of the complex nature of policy reforms and the uncertainty of their effects. The interrelationships between forests and other sectors of the economy are also better understood. Finally, there is a greater recognition that policy statements mean little in practice without strong institutional capacity to implement them.

Complexity and uncertain consequences of forest policy reforms

Debates focusing on policies for sustainable forest management have highlighted the complexity of cause and effect relationships and the uncertain consequences of policy reforms. During the past few years, policy analysts have focused on such issues as the underlying causes of deforestation, the effects of trade liberalization and the impact of structural adjustment policies, but the practical policy implications of these analyses are still far from clear. Their differing conclusions have highlighted the fact that the same policy measures can produce very different results depending on the national (and local) circumstances.

For example, there is intense argument about whether trade liberalization leads to better forest management and conservation or, instead, to forest depletion. Many environmentalists warn that trade liberalization policies facilitate the unrestricted operation of some transnational forest corporations that have a history of using economically and environmentally unsustainable and socially regressive methods of production. However, advocates of free trade argue that trade liberalization policies result in economic expansion and that – as demonstrated by the experience of industrialized countries – greater income levels lead eventually to more sustainable forest management.

Also, some analysts have pointed out that economic adjustment programmes promoted by international institutions have augmented pressure on forest resources. They contend that these programmes have often resulted in higher unemployment and the collapse of small business enterprises, which in turn lead to unsustainable use of forest resources, i.e. increased clearing of forests for agriculture and / or overharvesting of forest products, particularly fuelwood (as people fall back on it as a household energy source). Others observe that greater economic affluence resulting from economic adjustment leads to improved long-term environmental management and lower rates of deforestation and degradation.

These examples illustrate the great uncertainty about the consequences of national policies. It is extremely difficult to isolate the effects of individual policies.

Intersectoral policy linkages

The complexity of policy effects is leading to increased attempts to account for intersectoral linkages in the design of national forest policies. Some countries have accepted that, despite expected increases in agricultural productivity, additional forest lands will inevitably be converted to agricultural uses. Relatively limited attention, however, is accorded to the design of sound policies which would contribute to more effective land use transitions. The situation is similar with other sectors that have considerable impact on forests, such as mining, oil and natural gas exploitation, agribusiness expansion and infrastructure development. Attempts to integrate intersectoral considerations into policy design for greater efficiency and effectiveness are still comparatively weak. Similarly, policy reforms in other sectors generally still give little importance to spillover effects on the sustainability of forests. There have, however, been some initiatives to provide compensation in return for benefits provided by forests to other sectors. For example, Costa Rica has created a system of payment for hydrological services of forests, and in the United States, New York City has created markets for forests' watershed protection services. In other cases, pressure from environmental groups has forced policy-making bodies to give greater attention to these issues.

Policy reform and the changing public/private interface

Various governments have tried to deal with weaknesses in policy implementation by broadening the responsibilities of the private sector and by focusing on a few key areas of the interface between the public and private sectors. Some countries have launched initiatives to privatize control and management of forest resources and functions traditionally in the hands of the public sector. Many governments have also attempted to improve the interaction between the private and public sectors, mainly by eliminating some "perverse" subsidies, by tackling illegal and corrupt activities and by rationalizing forest concession policies.

Privatization. Some countries, disappointed with the poor results and inefficiencies of government agencies' management of public forest resources and forest-based companies, have taken some radical steps to privatize these resources. Colombia has recognized and assigned formal property rights to lands that have been held by rural populations under traditional land tenure systems. Honduras has experimented with the privatization of large areas of the country's pine forests and has sold various public forest-based companies to private interests. The United Kingdom and New Zealand have further consolidated their privatization programmes started in the 1980s (see Box 19). Other countries, such as Peru, have given serious consideration to the privatization of a number of forest areas and functions.

Perverse subsidies. Some countries have taken steps to eliminate or substantially reduce subsidies in the forest sector or in other sectors (e.g. transportation, agriculture and cattle ranching) that have had unintended negative effects on forests. These policy reforms have frequently been helped by the spread of fiscal austerity measures and in some circumstances by the reduction of international markets for the subsidized products. (Arguably, this was the case of policy reforms that eliminated subsidies to cattle ranching, which coincided with receding international markets for beef.)

Illegal and corrupt activities. Illegal and corrupt activities have acquired greater visibility in a number of countries. Brazil has formed a Presidential Commission to investigate the magnitude and effect of illegal and corrupt activities in the forest sector, with a view to reducing their incidence through policy reform. Malaysia has taken decisive steps to recapture revenues lost because some 20 unscrupulous corporations had actively engaged in transfer pricing. The World Bank has formally declared an attack on illegal activities and has begun to impose loan conditions to induce the creation of mechanisms for reducing the impact of corruption, based on independent third-party scrutiny.

Forest concessions. During the past few years Bolivia, Brazil, Cameroon, Canada, Costa Rica, Malaysia, Suriname and other countries have changed their forest concession policies in order to reduce or eliminate waste, foster more sustainable forest utilization and management and promote the well-being of local populations. New policies incorporate various mechanisms (which had been suggested by experts repeatedly in the past but rarely adopted in practice) for assessing concession values on the basis of timber volumes extracted rather than areas involved, for establishing performance bonds and for supporting third-party monitoring (see Box 20).

EMERGING TRENDS IN FOREST LEGISLATION[33]

As a corollary to the policy and institutional developments discussed above, recent years have witnessed a significant acceleration in the revision of forest-related laws around the world. Not

[33] This section is based primarily on a review of laws and law reform processes under way in Africa, Latin America, North America and central and eastern Europe. As used in this summary, the terms "law" and "legislation" generally refer to principal legislation as well as associated regulations and other subsidiary legal documents. "Forest law" should not be thought of as a stand-alone body of legislation that is specific to the forestry sector. Though most countries have principal forest laws, the legal framework governing forest resources is properly conceived of as comprising laws covering a wide range of issues of relevance to forests, from trade and taxation to environment, land use and agriculture.

BOX 19

Privatization of public forests in the United Kingdom and New Zealand

Both the United Kingdom and New Zealand have privatized significant portions of their State-owned forest resources since the early 1980s. The approaches taken to privatization differed between the two countries, with the United Kingdom pursuing a gradual disposal policy and the New Zealand Government choosing to sell most of its forest plantations over a short period.

United Kingdom

Privatization was introduced as a major policy initiative of the incoming Conservative government in 1979. Sale of State-owned forests started in the mid-1980s, with the intention of rationalizing the management of the State's 900 000-ha forest estate. Targets for the revenue from sales of forest and the area to be sold were set by the government (£150 million, or about US$315 million, and 100 000 ha by the year 2000).

Forests were sold by competitive tender or negotiation. By March 1997, the Forestry Commission had sold 66 000 ha and raised £75 million (about US$120 million). The forests sold were mostly the conifer plantations furthest away from district offices or those that in some other way were difficult to manage. Forests providing significant non-timber benefits generally were not sold. Nevertheless, concerns began to be raised in the early 1990s about the loss of public access to privatized forests – outdoor recreation being a major use of State-owned forests in the United Kingdom. In response, the Forestry Commission initiated a policy of giving local government the opportunity to enter into formal and legally binding access agreements for areas of forest about to be sold.

Complete privatization of the remaining State-owned forest estate was considered in 1994. This was rejected, however, on the grounds that: it was unlikely that the whole resource could be sold all at once for a reasonable amount; the sale would be legally and administratively complicated, and thus expensive; and the public would resist such a move.

The policy of gradual disposal therefore remained, with annual area targets similar to before. The new Labour government, however, has halted all sales for the time being.

New Zealand

The New Zealand Government announced in December 1987 its intention to privatize public assets that had commercial functions, including forests. The primary reason for selling government assets was to reduce public debt. The main rationale for forest asset sales, however, was probably ideological – namely, the belief that it was inappropriate for the State to own commercial forests. Only forest plantations, which supply most of the country's wood production, were sold; all publicly owned natural forests were transferred to the Department of Conservation. A series of sales took place between 1990 and 1996, resulting in the purchase of about 510 000 ha of forest plantations.

A major objective was to produce a more efficient, internationally competitive forest sector. A particular concern was the need to provide security of supplies to processors in order to attract new investment in forestry value-adding industries. The sale of forests to enable processors to integrate supplier functions into their current operations was seen as a means of achieving this end.

The process had two phases. First, government forest plantations (and processing facilities) were transferred to a government-owned corporation. Following the development of commercially viable operations, the assets were put up for sale. The purchaser bought the trees and fixed assets, but acquired only a lease on the forest land from the government under a tradable Crown Forestry Licence. The buyers included New Zealand forestry corporations, several Asian-based companies and an American company.

An important question in any privatization process is whether the sale should be open to foreign investors. In New Zealand the decision was relatively straightforward. Revenue maximization was most likely to be achieved by placing as few constraints on the process as possible and by maximizing bidding competition. The government recognized that overseas investors had potentially greater access to capital and were consequently more likely to invest in new processing facilities. This has generally proved to be the case. The entry of foreign investors into the forestry sector has had a number of positive effects: it has encouraged introduction of new technologies, improved market awareness and opportunities and increased efficiency through enhanced domestic competition. The major drawback was a negative public perception of foreign control of resources.

BOX 20
Pricing forest concessions

The government is the main owner of forest resources in most countries. Governments issue formal permits – forest concessions – to individuals or private corporations to manage certain areas of public forest for timber production. Forest concessions grant exclusive rights to assess forest production potential, to harvest wood or other forest products and to manage a specified area of public forests for a certain period. Here, the term is taken to be synonymous with "forest utilization contracts".

Harvesting of forests creates forest economic rents, which are returns from forest harvesting beyond those that could be earned by the best alternative use of productive resources elsewhere in the economy. Governments can capture part or all of the economic rent through a variety of forest charges, including taxes, charges per volume harvested, area charges and exploration fees. If these rents are not captured by government, they accrue to private concessionaires in the form of exceptional profits.

The amount of money paid to the government by the concessionaire through the various forest charges often bears little relationship to the market value of the resources and to the forest rents generated. This unsatisfactory situation is common in both developed and developing countries. As a result, during the past few years several forested countries have reformed their forest concession policies. These reforms should increase economic efficiency, improve forest management and lessen social and environmental impacts of concessions on public forest lands.

Large areas of forest are under concessions in some countries, and various companies are aggressively seeking new concession contracts in resource-rich countries. In Guyana, one company in 1991 obtained logging rights for almost 1.7 million hectares, or 8 percent of the country's land area. By 1997, about 30 percent of Guyana's forests were reported to be under concession contracts. More than half the closed forests in Indonesia are managed under timber concession schemes. In 1995, 2.4 million hectares of Suriname's forests, or 16 percent of the country's resources, were under concession management. Venezuela's forest concessions covered some 3 million hectares in 1995, but the five-year National Development Plan contemplates an expansion to 10 million hectares. Virtually all of Cambodia's forests, except those in national parks and protected areas, are now under forest concessions.

Various analyses have indicated that governments granting forest concessions often capture only a small proportion of the forest value. While the estimates (and the assumptions used to make the estimates) are frequently debated, evidence strongly suggests that in many cases governments capture less than 50 percent of forest rents, and often much less.

Why is this underpricing in forest concessions important? Underpricing encourages logging companies to use wood wastefully because it is inexpensive. Companies may obtain access to public forest areas that are far larger than they could otherwise afford, and they may rush to harvest only the most valuable wood. Inefficient wood processing companies survive because the raw material is inexpensive. Exceptional profits also create greater uncertainty about how long the concessionaire will be able to hold the profitable public concession, thus creating an incentive for quick and careless logging. In short, underpricing of wood reduces the concessionaire's incentive to invest in more sustainable forest management practices. Furthermore, and particularly in the case of developing countries that face a chronic shortage of funds, undercharging for timber concessions deprives governments of a source of revenue, thus making it more difficult for them to fulfil their responsibilities as stewards of the nation's forest wealth. Difficulties created by less than satisfactory rent capture become more serious in situations where structural adjustment policies have reduced forest department budgets. In these situations, the need to increase forest rent capture in timber concessions is more acute. Failure to capture economic rents also considerably weakens the case of forest departments in developing countries seeking increased financial support from the national treasury or from international donors.

Underpricing in forest concessions is one of the many "perverse" policy interventions that make sustainable management of forest resources difficult to achieve. Remedial actions in forest concessions must include much more than just getting prices right. Other dimensions of concession policies such as the type of charge introduced (area fees, volume charges, taxes, etc.), the length of concession contracts, procedures to avoid or minimize corruption and the degree of competition encouraged in forest markets are also important. However, establishing prices that would increase the government capture of economic rents is, without question, an important step for improved forest management and economic efficiency.

A number of countries are actively trying to introduce policy reforms to eliminate pricing and other policy distortions in forest concessions. In 1994, the Government of the Canadian Province of British Columbia, convinced that forest charges were far too low, nearly doubled stumpage prices. This led to a dramatic 120 percent increase in forest revenues, with no indication that less wood was harvested by private corporations in the two years after the stumpage price increase. In 1993, investors reportedly offered Guyana area fees that were 150 times established rates (Sizer, 1996). Cameroon, following advice from the World Bank, implemented a bidding process in 1995 to assign concessions. Bidding was expected to increase competition and produce prices closer to the economic value of timber. It did; government revenues rose more than fourfold between 1995 and 1996. These are clear indications that forest charges were far too low before the policy reforms were introduced.

Similar steps taken in Costa Rica, Ghana and Malaysia succeeded in dramatically increasing the government capture of economic rents from forest concessions. Guyana and Suriname have also attempted, with varying degrees of success, to stop indiscriminate granting of forest concessions at very low prices until procedures and institutions could be put in place to price and administer forest concession contracts effectively.

One may ask why such reforms have not been more common or have not been initiated sooner. In many cases, underpricing of wood arises from a desire to promote forest-based industrial development, to increase log exports or to make forest lands more accessible to poor rural populations. At least part of the reason for the persistence of underpriced wood in timber concessions lies in the political complexities of introducing reforms that would increase prices to concessionaires. The faulty forest concession systems that predominate in many forested countries of the world owe their existence and persistence to the influence of powerful vested interests that benefit from the perpetuation of the agreements. These interests exert their political muscle to prevent change. Another plausible reason in developing countries is related to the weakness of public forest administrations. Improved forest pricing in forest concessions requires the institutional capacity to assess values, to negotiate agreements and to monitor and impose implementation of the new rules. This institutional capacity is weak or non-existent in many developing countries. The fact that some developing countries are effectively implementing pricing reforms despite institutional weakness, however, indicates that when the political will exists, substantial improvements are feasible.

surprisingly, the results of these law reform efforts have been extremely varied. They have taken place within the context of vastly different legal and political traditions, reflecting a wide range of economic, ecological and social variables.

It is possible, nevertheless, to identify several trends that have achieved prominence over the last decade. Broadly speaking, forest law in the 1990s is moving away from a regulatory approach focused primarily on government management and policing of forests as economic resources. It increasingly recognizes the multiple interests involved in or affected by forest management, with greater attention given to the environmental and social roles of forest resources and a new emphasis on the involvement of a wider range of public and private actors.

This section reports on four areas in which a reorientation of national forest laws is evident: local forest management, the environmental functions of forests, forest management planning and the granting of forest utilization contracts.

Promoting local forest management

Local forest management has received a great deal of attention in recent forest law reforms, especially with respect to community-based activities and the realignment of power and responsibilities between central and local governments.

Historically, forest laws in many countries have

tended to be inhospitable to local forest management. Although older laws have often given recognition to limited usage rights, they have provided little scope for local people to play a meaningful part in the planning and management of forest resources on which they may have depended for generations. In many cases, the State has arrogated this role to itself through the creation of State forests. In other contexts, national law may have left the tenurial status of forest areas unclear, giving weak or no legal protection to existing community-based tenure systems and providing no alternative mechanisms by which local groups or individuals might assert effective control.

Efforts to address these shortcomings in recent legal changes have taken various forms.

First, there has been a proliferation of new mechanisms for the devolution of forest management to local communities, user groups or households through site-specific arrangements such as co-management agreements, community forestry leases, the delineation and titling of village territories and related devices. Nepal's 1993 Forest Act offers a notable example of this approach, providing for the turning over of portions of national forest to local user groups which agree to manage the areas in accordance with an approved plan. A range of variations on this approach may be found in recent laws or regulations in a growing number of countries or jurisdictions, including (to name only a few) Cameroon, Guinea, India, Laos, Madagascar, the Philippines, South Africa, the United Republic of Tanzania, the Province of British Columbia in Canada and a number of countries in Latin America.

Second, some countries have accorded increasing recognition to the historical land or territorial claims of local peoples. The 1997 Indigenous Peoples' Rights Act in the Philippines is an important recent example of this phenomenon. The rights of indigenous communities figure prominently in several Latin American laws. A number of other countries, including Australia, Canada, South Africa and several countries in central and eastern Europe, are engaged in restoring the lands of dispossessed communities and individuals, which may include natural forests or commercial plantations.

Third, as discussed in detail in the previous section, many countries have moved to decentralize various aspects of forest administration to local government bodies. This change may be reflected either specifically within forest legislation or more generally as part of an overall revision of local government law. Again, examples can be found in all regions, from the Sahel to Southeast Asia to Latin America.

These three categories are indicative only; they do not capture the full variety of approaches being tried throughout the world. Moreover, they are not mutually exclusive, and often overlap or operate concurrently. For example, in the case of several West African laws, where management of some forests has been decentralized to the level of the commune, the local government may have the opportunity further to devolve management over specific subareas through a contract with local users.

While there is unmistakably a higher profile given to local forest management in recent legislation, many of the reforms are characterized by significant limitations and ambivalence, both on paper and in practice. Forest authorities in some countries, for example, have embraced the concept of community-based management mainly in areas where forest resources are already depleted; there is frequently a reluctance to share control over richer, more intact areas, even where there is evidence that the health of the resource is due in large part to the long-established informal stewardship of local people. In many co-management schemes, government forest departments continue to retain most of the important decision-making power, including the power to draft and approve management plans and to decide about the selection of species, the marketing of harvested products and the use of benefits by the contracting group. The long-term security of the rights granted or recognized under local arrangements may be unclear, either because the term of the agreement is very short or because the government has wide power to terminate the agreement for poorly defined reasons. And in some instances, legal provisions that appear to protect the traditional rights of indigenous peoples may be so weakly or ambiguously drafted as to be unusable.

In short, there is a continuing need to find workable legal arrangements that provide or recognize meaningful and secure rights for local forest managers while allowing a flexible response to vastly different local conditions and an appropriate balance between various interests. It is very likely, therefore, that this will be an area of law that continues to experience significant evolution in the years ahead.

Enhancing the environmental functions of forests

Environmental concerns have never been absent from forest legislation. Nevertheless, particularly in the aftermath of UNCED, national forest laws now more explicitly reflect the environmental importance of forests than they did in the past. This concern is prominently articulated in the preamble or statement of objectives of many new laws, which now routinely refer to environmental objectives of forest management and to the country's international environmental obligations.

How these general goals are given specific legal content varies greatly from country to country. A variety of techniques is used in recent legislation:

- ***Inventory, planning and classification provisions.*** In recent laws these are often strongly tied to environmental considerations. For example, a requirement to collect and collate information concerning biological diversity and other environmental attributes may be included in legally mandated forest inventory provisions. Laws are also increasingly specific in prescribing that environmental criteria be taken into account in the drafting of forest management plans (see below). In addition, forest laws are now more likely to contain forest classification categories that explicitly embody environmental objectives, and to require that a forest under a given classification must be managed according to certain principles. Examples from recent forest laws are forest nature reserves, watershed forests and other categories designed to protect certain forests from commercial logging.
- ***Environmental impact assessment.*** This a related tool that is finding its way into forest management, either by explicit reference in forest legislation or in more general environmental legislation that would cover activities in forest areas. In some cases, the holders of large-scale concessions are required to prepare environmental management plans as a condition of their contract. In Bolivia, a 1996 act provides for forest inspections and audits, which among other things would periodically assess the environmental effects of forestry activities. An interesting feature of this legislation is that it allows community forestry groups to substitute a certification finding from an internationally accredited certification programme in place of the otherwise mandated audit (Markopoulos, 1998). (To date, certification has not figured prominently in forest legislation, although it is referred to in South Africa's National Forests Bill and in recent proposals in a number of other countries. It would not be surprising to see increasing references to certification in future legislative drafting.)

The increasingly environmental slant of forest legislation, coupled with the emergence of environmental law in many countries around the world, has led in some cases to significant, if avoidable, legislative confusion. Forest laws and general environmental laws are not infrequently drafted with overlapping and conflicting provisions, reflecting unclear mandates and institutional jealousies. As forest law continues to develop in scope and complexity, it is increasingly important for lawmakers to strive to calibrate its provisions more precisely with related bodies of law.

Management planning

Forest management plans have long been a feature of forest laws. Nevertheless, there is a discernible trend towards more widespread and detailed attention to planning in forest legislation and a broadening of the objectives of the planning process. Laws increasingly require planners to consider a wide range of ecological and social issues that might previously have been outside the purview of traditional forest management planning. In addition, a hierarchy of plans may be called for, with plans for local management units required to be consistent with an umbrella regional or national plan.

Planning procedures in recent forest laws can also provide an important point of intervention for public participation in the design of forest management. It is increasingly common for laws to require public notice at various stages during the planning process, opportunities to comment, public meetings and access to preliminary drafts of plans.

It is difficult to generalize about the legal status accorded to management plans. In government forest contexts, where a forest department is responsible for producing a plan, laws have often been silent about the plan's legal effect. Nevertheless, even in such cases a number of forest laws now specifically require that administrative action on the part of forestry officials – such as the granting of licences or other permissions – must be consistent with an approved plan. In other situations, such as timber concessions or community forestry agreements, the plan may in essence be considered part of the contract itself.

Despite the prominence given to management plans in legislation, their application on the ground has tended to lag far behind in many countries. In this area, there is often a disparity between the vision of those drafting the law and the capacity of government or private forest managers to implement it. Part of the problem may be in terminology, where reference to a "management plan" conjures up a more complicated and expensive instrument than is feasible or desirable to prepare in a particular context. Cameroon deals with this issue by requiring relatively simple plans in community or private forestry.

Granting of forest utilization contracts

The importance of sound policies concerning the pricing and award of concessions and other forest utilization contracts has been discussed in detail above. From a legislative perspective, many older forest laws and regulations are remarkably silent as to what procedures and criteria should govern the awarding of forestry contracts. As a result, in many parts of the world the granting of forestry concessions tends to be a secretive affair, often conducted at high political levels on an ad hoc basis. Not infrequently foresters find that concessions are presented to them as a *fait accompli* by their political superiors, without technical criteria taken into account.

The general trend in recent years has been in the direction of spelling out in some detail the steps leading up to the awarding of a contract, sometimes in principal legislation, sometimes in regulations, or most often in a combination of the two. Such legal frameworks set forth, for example, the basic mechanical elements of an auction and bidding system, such as the content of the call for bids, the form and content of submissions, deadlines and decision-making time frames and the professional qualifications and independence of auctioneers. Laws may also specify when government can use non-economic criteria to distinguish among competing bids. A country may, for example, want to give weight to certain non-quantifiable characteristics of the bidding entity; for instance, the Bolivia law of 1996 gives certain preferences in concession awards to indigenous people in the vicinity.

Specifying the process of awarding contracts may be helpful in a number of ways. It promotes greater transparency and accountability in decision-making and helps provide a level playing field for potential contractors. It can also help ensure that technical specialists are involved in the decision-making process, as opposed to simply finding decisions foisted on them. Finally, a well-ordered, consistent process of evaluating and granting contracts may better promote the interests of third parties and the public at large. It can contribute to reducing the chaos and conflict that has emerged in some countries where carelessness or worse has led to the issuing of overlapping concessions or to the partial or complete disregard of existing rights within a concession area.

Putting legislation into practice

The purpose of this brief survey has been to provide a sampling of recent developments in national forest laws. It is, of course, not exhaustive in either geographic coverage (the emphasis has been on developing countries and countries in transition) or the issues discussed. As is also evident, the focus here has been on written law. There are of course often great discrepancies between the content of law and its implementation on the ground. Any number of factors may contribute to this discrepancy – insufficient political will, weak institutional capacities, overburdened court systems, the absence of suitable incentives for participation or compliance by stakeholders, financial difficulties, corruption, etc. These factors do not lessen the importance of establishing sound and coherent legal frameworks. They do serve as a reminder, however, that law reform should, above all, be an exercise in realism. Laws that rely on resources that governments do not have or that require abrupt reorientations of institutional or social behaviour are likely to fail. Effective legal reform requires an assessment of the capacity and willingness of government and others to implement new legal strategies, a commitment to monitoring the effects of legal changes over time and a responsiveness to the results of that monitoring. ◆

Evolving institutional framework

Changes in national economic policies and societal demands on forests are having direct and substantial impacts on government forest institutions and administrative arrangements for forest management. These include modifications in the role of forest administrations, a move towards decentralization and changes in forest research and education orientations and structures.

NEW ROLE AND STRUCTURE FOR FOREST ADMINISTRATIONS

In many countries, central forest administrations have moved away from their traditional role as supreme forest guardians. In some places, the normative functions of policy formulation have been separated from operational responsibilities, which increasingly have been passed to the private sector, non-governmental organizations (NGOs) and local groups.

This deliberate shift of responsibilities away from centralized public management has taken four main forms:

- In the most extreme form of privatization, government has transferred significant areas of forest and public forest companies to the private sector (e.g. New Zealand and the United Kingdom – see Box 19, above), relinquishing to a great extent both normative and operational responsibilities. It has generally been production forests or forests with lower social or environmental value that have been privatized. Unprofitable and inefficient public companies have also been privatized.
- In a second scenario, government has kept the normative responsibility of deciding how forest resources should be managed, but has entrusted the private sector with implementing government-designed plans, as in the case of contract logging.
- Central government has in other cases entrusted to local governments and local communities the responsibility of designing and implementing forest management strategies and plans (see section on decentralization which follows).
- Some governments (e.g. Ghana, Papua New Guinea, Suriname, Uganda, Zambia and Zimbabwe) have experimented with the establishment of semi-autonomous entities and private-public joint ventures. Forestry administrations in Indonesia, Malaysia and Uganda have developed partnerships with the private sector and NGOs in such areas as research, extension, revenue collection and policy enforcement.

Many industrialized countries have moved substantially towards privatization and decentralization. The shift in public forestry administrations' role has been more gradual in developing countries, where necessary policy and legislative reform and implementation have frequently been subject to delays, and where the historical context is different.

Moves towards decentralization have coincided with increased efforts by public administrations to involve a range of interest groups in public debate about forests and in decision-making processes. The forestry sector today is characterized by a pluralistic or multistakeholder environment in which an increasing number of independent groups (NGOs, different levels and branches of government, rural people's organizations, private-sector concerns, political parties, unions, etc.) are demanding, and often obtaining, a greater role in decision-making about forests, forestry and rural development.

These groups often have widely differing opinions, objectives and knowledge about forestry. Traditional decision-making mechanisms are being strained by the conflicting interests. Some mechanisms even seem on the verge of paralysis, as witnessed by the large number of forest management plans stalled in the courts in the

United States and resistance to decisions and actions by forestry departments in many countries.

The existence of multiple and competing interests undoubtedly adds to the richness of the forestry debate. It provides an institutional system of checks and balances, assuring that decisions are closely monitored and questioned. The question is not whether to accommodate multiple interests but how. Without coordination and collaborative mechanisms, potential fragmentation of organizational responsibility poses a clear risk to sustainable forestry and rural development. Many efforts in eliciting public participation and using participatory approaches have not been very successful, partly because they have sometimes been processes of co-option and "rubber-stamping" instead of negotiation and dialogue. Methods and tools developed for participatory approaches and conflict management can be useful in developing collaborative mechanisms (see Box 21). Communication and learning seem to be at the centre of any attempt at coordination in pluralistic environments.

The role of forest institutions as facilitators among a range of stakeholders has become increasingly important. This is evident, for instance, in North America and Scandinavia, where forest institutions are becoming brokers among competing positions on best approaches to forest resource utilization. Conflict management techniques are increasingly being used in the forest sector, at both the local and national levels and in developed and developing countries alike (see Box 22).

EFFECTS OF DECENTRALIZATION ON FOREST MANAGEMENT

Countries' efforts to decentralize certain functions of central government administrations have included changes in the forestry sector. In West Africa the process has taken off over the past few years, particularly in French-speaking countries such as Burkina Faso, Mali and Senegal. Bolivia and the Philippines are also often cited as countries actively promoting greater involvement of local government in natural resource management.

A number of factors are likely to support

BOX 21
Tools and fora for accommodating multiple interests in forestry

Tools and methods have been developed to deal with multistakeholder situations. They include:

- ***collaborative learning:*** techniques for building agreement using tools from conflict management and multidisciplinary systems methodologies, which have been used to plan and implement forest fire recovery programmes in the United States;
- ***patrimonial mediation and management subsidiarity:*** mechanisms for reaching agreement on historical processes and future scenarios and for promoting decentralization of decision-making to the lowest competent level, which are being used in forestry programmes in Madagascar;
- ***co-management of protected areas:*** collaborative management which combines participatory methods with distribution of responsibility and formal vesting of authority in local groups and communities, as in programmes supported by the World Conservation Union (IUCN);
- ***adaptive management of renewable resources:*** management that consciously promotes continuous learning and adaptation, as has been used in Queensland, Australia and the Columbia Basin of the United States and Canada, among other places.

These methods, although still evolving and imperfect, reflect a growing recognition of the necessity of incorporating and influencing social processes in forestry. They are all multistage and iterative processes, with an emphasis on learning and dialogue. They also accept some degree of power sharing among interest groups.

Various fora at all levels are providing increased opportunities for groups to interact and negotiate. These include a whole range of local processes as well as national forest programmes, the Intergovernmental Forum on Forests, model forest programmes and the World Commission on Forests and Sustainable Development (see Part III).

Despite the growing set of means for assessing multiple interests, strengthening cooperation and developing collaborative mechanisms, further research and development are still needed. Accommodating multiple interests may be complicated and will not always be successful, but it appears to be unavoidable.

BOX 22
Natural resource conflict management

Many countries and institutions recognize the need to understand the causes and consequences of conflicts over the ownership, management and use of natural resources, including forests and trees, and to incorporate conflict management mechanisms into natural resource planning and management. Conflict management is based on a dialogue leading to a mutually acceptable solution. Institutions around the world are focusing on capacity building, community empowerment and policy and legislative reform in support of this approach.

In Asia, for example, the Regional Community Forestry Training Centre (RECOFTC) in Bangkok, Thailand is developing a conflict resolution training manual series, which covers basic resolution techniques as well as intervention skills for dealing with environmental, economic, gender and cultural conflicts. Other institutions working on capacity building for conflict management are Environment and Development in the Third World Sahel and West Africa's Forum de groupes de recherche action formative (ENDA GRAF Sahel), working with national networks in West Africa, and the Universidad de la Paz in Costa Rica and the Universidad Politécnica Salesiana in Ecuador which have developed training programmes.

It is necessary to reduce power disparities before parties can engage in effective dialogue. Women and marginalized groups in particular are often disadvantaged in this respect. In Ecuador, Bolivia and Peru, a working group involving the Instituto Latinoamericano de Investigaciones Sociales (ILDIS), IUCN, FAO's Forests, Trees and People Programme (FTPP) and other institutions has helped indigenous communities strengthen their positions and negotiating skills in conflicts with oil companies over petroleum exploitation in forest areas. In Ecuador, the process has resulted in the creation of an Environmental Technical Committee with representatives of the indigenous peoples, the oil company and the government.

National laws that recognize and support community-based forest management and property rights, such as the newly enacted Indigenous Peoples' Rights Act of 1997 in the Philippines, provide a means for more equitable processing of conflicts. Moreover, conflict management techniques can be used to formulate policy in a participatory manner. For example, the Fundación Futuro Latinoamericano (FFLA), a private organization based in Ecuador and specialized in conflict resolution and participatory policy review processes, facilitated negotiation of the revision of Bolivia's forestry legislation.

continued decentralization efforts. These include some factors specific to the forest sector:

- the realization that in many countries forest protection and management have not been very effective when centrally controlled;
- the need for clear separation of functions and systematic organizational checks and balances;
- the inability of governments at the central level to address adequately the growing number of conflicts over forest resources at the local level;
- reduced budgets and staffing levels in forest services at the central level.

Other factors are related to political processes (empowerment and participation):

- increased recognition of the rights of indigenous people;
- the realization that in order to achieve effective forest management, all stakeholders need to be involved in the decision-making process;
- a desire to promote more equitable sharing of benefits from forest resources;
- growing pressure for greater access to, and control over, forest resources by local entities (local authorities, community-based groups, forest user groups, etc.).

Others are related to national or international priorities:

- the desire to promote more balanced socio-economic development in a national context;
- increased pressure by donor countries for countries to engage in decentralization processes.

The impact of decentralization on natural resource management is difficult to assess, partly because it has been initiated only relatively recently. In addition, the decentralization process differs from country to country and from sector to sector, which makes comparison and evaluation between countries, or even between sectors in the same country, difficult. Analysis is also hindered by the variability in capacity and resources of the lower levels of government, communities and newly created governance entities to whom the power and responsibility are given.

Although it may still be too early to conduct a thorough assessment of the impact of decentralization on forest management, certain lessons are emerging. Often, when authority and responsibility have been shifted to the local level, additional human and financial resources commensurate with the new responsibilities have not been provided. This was the case with the Integrated Social Forestry (ISF) programme in the Philippines. The Department of Environment and Natural Resources, prior to the Local Governance Code of 1991, recommended that the ISF programme be decentralized so as to be more responsive to diverse local needs and conditions. When decentralization was officially implemented, however, not all the provinces received their share of personnel, materials and equipment. Without adequate human, financial and technical resources, local governments had difficulties coping with the volume and demands of ISF activities. In addition, local governments assigned low priority to ISF, as with other non-revenue generating projects. Several years after the Local Governance Code had been enacted, it was found that the performance of ISF projects had declined.

This example and experiences elsewhere, such as in Bolivia (see Box 23), illustrate that certain conditions need to be met for decentralization to have a positive impact on forest management. These include:

- sufficient transfer of authority, in addition to responsibility, from central forestry agencies to the local level;
- sufficient financial resources and mechanisms for generation of local revenue accompanying the transfer of responsibilities (i.e. fiscal decentralization);

BOX 23
Decentralization: the example of Bolivia

With the expectation that decentralization could lead to more efficient natural resource management, several Latin American countries have recently transferred rights and responsibilities for forest management and conservation to municipal and lower-level local governments. Nowhere has this process gone as far as in Bolivia. The country's 1994 Popular Participation Law devolved a broad range of responsibilities to municipal governments, including education, health and urban infrastructure, and guaranteed them a certain percentage of the national budget. The 1996 Forestry Law allocated 20 percent of public forests to municipal administrations for use by community groups and gave local governments direct control over 25 percent of the royalties from these forest concessions.

The impacts of these changes so far have been mixed. Previously marginalized groups, such as indigenous people, small farmers and small-scale timber producers, have benefited from increased access to forest resources and from greater opportunity to influence municipal governments on decisions related to local forest management. Several municipalities have established municipal forestry units, and local governments have become more involved in a wide variety of activities related to forest management, agroforestry, protected areas and land use planning.

Not all local groups, however, have been equally well positioned to take advantage of the opportunities, and improvements in forest management have not always been achieved. Important problems remain, such as weak technical and managerial skills within both local governments and communities, and the lack of necessary external support, particularly from national and departmental governments, to strengthen local capacities. The overall policy context is insufficient to allow local groups to exercise their legal rights and responsibilities fully, and as a result has tended to stymie local forest management initiatives.

Decentralization has sometimes given local communities the means needed to protect their resources well, while on other occasions it has enabled the misuse of power by local groups and individuals which has sometimes led to forest degradation.

- adequate administrative and technical capacity for handling new responsibilities;
- reliable accountability mechanisms in place;
- a clear definition of what should be transferred and how, as well as to whom the transfer should be made.

With growing momentum for decentralization it is anticipated that local communities will have better opportunities for negotiating a greater role in managing and benefiting from local forest resources. The process is expected to allow local governments to be more responsive and accountable to the needs of local stakeholders. The underlying assumption is that local institutions will have a greater sense of responsibility for stewardship of forest resources, in part because this will be linked to a greater share of the benefits. Many cases today, such as the forest user groups in Nepal, illustrate this point. Decentralization of management responsibilities to local groups has resulted in rehabilitation of degraded lands, planting of new forests and improved forest management efforts.

While decentralization is showing many positive signs of leading to more efficient and sustainable forest management, particularly by providing more opportunities for local involvement in decision-making, it carries potential risks. When rights and responsibilities for forest management are transferred to the district or municipal level, central government control and support are often reduced. The potential exists for the local élite or special interest groups to gain control of the management of forest resources. One way to guard against this result is to ensure that where representative, equitable and accountable traditional systems of local governance exist, they are included in decision-making on management of forest resources. It is also important that adequate local capacity, information and trained personnel be in place to ensure that new responsibilities are effectively met at the local level.

Moreover, clear enabling policies are needed, and legislative systems and institutions must be coordinated, to provide a supportive and well-understood framework to achieve sustainable forest management and to create incentives at the local level to harmonize development and conservation efforts. As already indicated, mechanisms must be put in place to ensure that adequate financial resources are made available to responsible decentralized institutions and that financial returns from forestry activities are equitably distributed. An adequate central regulatory structure must also be in place to ensure that laws and regulations are followed.

Although it is still early for full assessment of the impact of decentralization on management of forests and other natural resources, experience has shown that when greater decision-making power and more secure rights are transferred to the local level, with clear definition of roles and responsibilities, natural resource management tends to be more sustainable and effective.

TRENDS IN FORESTRY RESEARCH AND TECHNOLOGY TRANSFER

Forestry research is being called upon to address the needs of a larger array of users in more technical fields than ever before. Various institutional changes, however, are affecting the ability of forestry research in both developing and developed countries to meet these needs.

In many developing countries, research capability is inadequately developed, existing research institutions are weak, access to information is poor and research tends to be fragmented and narrowly focused on traditional topics. Strengthening national forest research systems is more than ever an important priority. Various economic and policy reforms, including structural adjustment, decentralization and privatization, however, risk having the opposite effect.

Ongoing structural adjustment efforts in many sub-Saharan African countries, as elsewhere, have led to the downsizing of public institutions. National agricultural research systems (NARS), of which forestry research is a component, have also been restructured. The prevailing trend has been to decentralize agricultural research by transferring national research staff to regional multidisciplinary research programmes within the country. While justified in principle, this decentralization may in fact weaken national capacity in forestry research, at least over the short term, by diluting a body of expertise already below the critical mass.

One way in which development projects have tried to compensate for weak national research capabilities is to include a research component in their activities. Although this has an invigorating effect on the particular field of research during the project's life, the benefits are unsustainable if permanent national forest research institutions are unable to participate in the activities and to capitalize on the results, as is often the case.

New institutional arrangements in forestry research are emerging in some developing countries to help compensate for reduced government support (FAO, 1997k; FAO/FORSPA, 1998). Universities, which have been found to be important but underutilized reservoirs of expertise (Kowero and Spilsbury, 1997), NGOs and the private sector are becoming more active in forestry research. While these developments are positive in principle, whether they will result in greater research efficiency and effectiveness will depend on good coordination among these groups and the ability to capitalize on their relative strengths and expertise. Another significant change is that research programmes are being forced to be more accountable to end users as users are becoming better organized and as governments increasingly insist on cost recovery or cost-sharing approaches.

Research is being privatized and decentralized in some industrialized countries as well. The results have been mixed. Decentralization often leads to decreased support for research because regional and provincial governments tend to assign low priority to long-term activities. Privatization of research in some countries, and the private sector's involvement in an increasing proportion of research in others, is resulting in a shift towards adaptive research to meet immediate and commercially oriented needs. Partnerships between private companies and universities are also changing the nature of university research programmes from basic to more applied research. While a more needs-oriented targeting of research effort is a positive development, the risk is that insufficient support will be given to basic and long-term scientific research (needed as a foundation for applied research) and to research addressing social and environmental needs.

At the international level, the forestry and agroforestry research programmes of the Center for International Forestry Research (CIFOR), the International Centre for Research in Agroforestry (ICRAF) and the International Plant Genetic Resources Institute (IPGRI) continue to develop steadily, while programmes of some other international agricultural research centres (IARCs) of the Consultative Group on International Agricultural Research (CGIAR) system have experienced major cuts. CIFOR, ICRAF and IPGRI, by mandate, focus on strategic and applied research at the global level, which they carry out in collaboration with regional and national research institutions. The results need to be adapted to local conditions by the national research systems. Weak research capacity at the national level, however, is recognized by the IARCs as an important constraint to widespread application of their research results.

An important effort to mobilize the world scientific community around agricultural research and in turn to help strengthen national research capacity in developing countries resulted in the establishment of the Global Forum on Agricultural Research (GFAR) in 1996. GFAR provides a framework for partnership between IARCs, research institutes (both public and private) in industrialized countries and NARS in developing countries. NARS are represented in GFAR by regional or subregional fora on agricultural research. Although in principle these fora cover forestry research, they are still concerned almost exclusively with agricultural research in the narrow sense. GFAR could help strengthen forestry research at all levels, but only if adequate attention is given to the forest and environment sector.

Although national, regional and international fora continue to discuss what research needs to do (i.e. priorities for research), questions on how to do it (i.e. institutional arrangements) have become more important in recent years. Increasingly, research institutions are involving research users in phases from programming through implementation of research and are developing improved methodologies to ensure effective and accelerated transfer of research results.

A major evolution in thought about technology transfer has occurred over the past decade. It is more commonly recognized that the solutions for problems are not always technical. Not every constraint to sustainable forest management can be alleviated by the transfer of technology. Policy environments favourable to sustainable forest management and to the implementation of technologies are as important as the technologies themselves.

Transfer of technology is no longer seen as unidirectional, e.g. from developed to developing countries or from technical institutions to user communities. Means of encouraging lateral and even upward transfer of technology and opportunities for more South-South exchange of information are increasingly being sought. In some cases, participatory technology development (based on the active participation of recipients in development and adaptation of technology) may be more effective and sustainable than the transfer of technology developed elsewhere. The value of traditional forest-related knowledge is also becoming more widely recognized. Significant opportunities exist for transfer of this traditional knowledge and environmentally sound local technologies. However, the lack of adequate protection of intellectual property rights and thus of the means to reward the supplier of knowledge is an important concern.

Participatory and problem-solving extension systems which accommodate the skills and knowledge of many partners are seen as essential to technology and human resource development for responsible forest management. In many countries, however, such systems for forestry are just emerging.

The Intergovernmental Panel on Forests developed proposals for action to improve research and transfer of technology at the global and regional levels. Further discussion is being undertaken by the Intergovernmental Forum on Forests. The second session of IFF (Geneva, Switzerland, August to September 1998) addressed questions of transfer of environmentally sound technologies. Issues that were stressed included partnerships among the public and private sectors, South-South collaboration, enabling policy environments, financing opportunities and both indigenous and modern technologies. Discussion of research issues is scheduled for IFF's third session (Geneva, May 1999). The points for consideration will be based on the results of the International Consultation on Research and Information Systems, an intersessional meeting held in Ort-Gmunden, Austria in September 1998. Issues addressed at this meeting included prioritization of research topics, capacity building, financing, global and regional research networks and global research systems.

The years to come will be crucial for forestry research and for the effectiveness with which the results are used in development. The two major current international initiatives – the establishment of GFAR and the international dialogue on forests led by IFF – can significantly influence the direction of forestry research and development. The great challenges will be to ensure that mechanisms are put in place for strengthening research capacities at all levels; that technology transfer issues are addressed; that research and technology transfer are driven by need and not by supply; and that international support is focused not on areas of research perceived as "politically correct" but on research required to facilitate long-term capacity development.

PREPARING FORESTERS FOR CHANGE: ADAPTING EDUCATION AND TRAINING PROGRAMMES TO NEW DEMANDS

Traditional forestry education has continued to be under pressure to evolve in response to new emphases in forest management and to changes in the job market. Educational institutions are expected to produce forestry graduates with a wider array of knowledge and skills than ever before. The complex and dynamic nature of sustainable forest management requires forest managers who have an expanded set of skills, are adaptable and are responsive to changing situations. The private sector and NGOs, which are increasingly important employers in the forestry sector, are looking for graduates with specific skills that were less needed in the past. Curriculum reform, expanded continuing education opportunities and a reconceptualization of the learning environment are some of the responses to the new demands.

Curriculum reform is probably the most obvious change in forestry education. A 1996 survey showed that more than 200 of the 750 forestry education and training institutions surveyed revised their curricula during the period 1989 to 1995 (FAO, 1996d). Revised forestry curricula tend to place increased emphasis on such subjects as ecology, environmental sciences, natural resource management, community forestry, agroforestry, marketing, management and administration. Some of the curriculum reform incorporates new techniques, learning models and styles, such as interdisciplinary problem-solving approaches, in addition to new content. In Malaysia, curricula were revised as part of a redesigning effort for forestry education which included the development of a strategic plan and the restructuring of faculty and departments. In the Czech Republic, as in other countries in transition, forestry curricula are being modified in response to dramatic shifts in forest ownership patterns and other changes brought about by economic and political restructuring. The Czech Republic's education system, which was formerly nationally uniform, has been diversified and adjusted to individual regions so as to be more applicable to their respective circumstances.

Meaningful curriculum revision remains difficult, however. Traditional but still important subjects, such as harvesting and transport systems, must still be taught while new subjects are added. Difficult trade-offs are necessary. In a survey of its graduates conducted by the Tropical Agriculture Research and Higher Education Center (CATIE), 96 percent thought that the profession's future lay in the private sector, while 91 percent stated that little training was received in management and administration – skills in demand by the private sector.

Adult learning and continuing education programmes have become increasingly important as a mechanism to fulfil demands for new skills. Employers of new graduates may need to provide preservice training programmes on specific subjects not covered by their formal training. There is also a growing reliance on continuing education or in-service training to give mid-career personnel new knowledge and skills.

Opportunities for learning outside the classroom are also increasing. Distance learning provides academic structure but flexibility as to place and time of learning. New information and communication technologies, such as the Internet, provide immediate access to unprecedented quantities of information and education resources. They are contributing to a reassessment of time and space in education, through concepts such as institutions without walls and learning as a lifelong endeavour. These technologies and their rapid pace of evolution, as well as other changes, are contributing to an inversion of the traditional learning pyramid, older professionals having to look to younger ones for information, training and advice.

Significant changes are coming about as a result of both decreased public support for forestry education and training and increased involvement of the private sector and NGOs in these activities. New strategic alliances between public and private institutions are emerging in some places. ForKom, a forestry forum in Indonesia involving private forest companies, government agencies, the Forestry Training College and international donors, is an example of informal coordination for human resource development in the forestry sector. In some places, private companies are co-sponsoring courses with universities in subjects of interest to them.

Despite the positive developments indicated above, overall education has been slow to adapt to new demands and far more change is still needed. Traditional institutions will be forced to explore ways of being more flexible such as allowing course work from a broader range of disciplines and giving credit for work experience and training at different institutions. Continuing education and in-service training programmes will have to be more frequent and will have to be designed and implemented better. Adult education and training approaches that will help foresters continue to learn and to update their skills will be increasingly important. Dynamic partnerships and alliances among a range of organizations – private sector, public sector and non-governmental – in development of training and education programmes, while becoming increasingly common, will require encouragement. ◆

PART III
THE INTERNATIONAL DIALOGUE AND INITIATIVES ON FORESTS

UNCED follow-up: an update of the IPF/IFF process

Forests were among the most controversial issues considered at the United Nations Conference on Environment and Development (UNCED) in Rio de Janeiro, Brazil in June 1992. The prevailing North-South polarization concerning forests did not permit agreements beyond the text of the "Non-Legally Binding Authoritative Statement of Principles for a Global Consensus on the Management, Conservation and Sustainable Development of All Types of Forests" (the so-called "Forest Principles") and Chapter 11 of Agenda 21 ("Combating deforestation").

By contrast, the "post-Rio" period 1992 to 1995 was one of confidence building and emerging North-South partnerships. This climate enabled the United Nations Commission on Sustainable Development (CSD) to establish the Intergovernmental Panel on Forests (IPF) at its third session in April 1995. IPF was set up for a two-year period and tasked with furthering the initiatives started at UNCED and encouraging international consensus on key issues related to forests.

By the time IPF completed its work in February 1997, it had developed over 100 negotiated proposals for action on issues related to sustainable forest management, including national forest programmes; forest assessment; criteria and indicators; traditional forest-related knowledge; and underlying causes of deforestation. These proposals for action have received worldwide recognition and endorsement by a large number of intergovernmental and professional fora, including: the fifth session of CSD, Earth Summit +5 (the Special Session of the General Assembly to Review and Appraise the Implementation of Agenda 21 – UNGASS); the United Nations Economic and Social Council (ECOSOC); the 1997 session of FAO's Committee on Forestry (COFO); the International Tropical Timber Council (ITTC); the fourth Conference of the Parties to the Convention on Biological Diversity; the meeting of the Commonwealth Heads of Government in Edinburgh, United Kingdom 1997; the G8 Summit in Birmingham, United Kingdom; and the third Ministerial Conference on the Protection of Forests in Europe (see below for more details).

Other indirect effects of the IPF process included:

- increasingly consistent instructions to intergovernmental organizations from their governing bodies on their forest-related activities;
- strengthened international networks among foresters, including English-, French- and Portuguese-speaking foresters' associations;
- increased attention to the valuation of environmental services of forests, as illustrated by the recognition of the significant role of forests in mitigating global climate change by the Kyoto Protocol of the Framework Convention on Climate Change (see below);
- review by many countries (e.g. Australia, Canada, Costa Rica, Denmark and Switzerland, as well as the co-sponsors of the Six-Country Initiative[34]) of their forest policies and national forest programmes in the context of IPF's conclusions and proposals for action.

The outcome of IPF was endorsed by the fifth session of CSD in April 1997. Various matters, however, were left pending, including issues related to finance and transfer of technology, trade and environment, institutions and legal

[34] The Six-Country Initiative was an effort made in 1997-1998 by Finland, Germany, Indonesia, Honduras, Uganda and the United Kingdom to review the implementation of IPF proposals for action at the national level. The initiative culminated at the International Expert Meeting on Putting the IPF Proposals for Action into Practice at the National Level, held in Baden-Baden, Germany in July 1998. The national case studies prepared and the results of the Baden-Baden meeting have been noted by the second session of IFF.

instruments. In view of these remaining issues, UNGASS, held in June 1997, recommended a continuation of the intergovernmental policy dialogue on forests. Subsequently, ECOSOC, at its meeting in Geneva in July 1997, decided to establish the ad hoc open-ended Intergovernmental Forum on Forests (IFF) under CSD. IFF will submit a progress report to the seventh session of CSD in 1999 and its final report to the eighth session of CSD in 2000.

IFF's first session (October 1997) established its programme of work, which falls into three main categories:

- implementation of IPF's proposals for action;
- addressing issues on which international consensus has yet to be achieved, including finance and technology transfer, trade and environment, underlying causes of deforestation and forest degradation, rehabilitation of forest cover and conservation, traditional forest-related knowledge, valuation of forests and the use of economic instruments;
- identification of possible elements of, and work towards, consensus on international arrangements and mechanisms for the management, conservation and sustainable development of forests, such as a legally binding instrument on all types of forests (see section on issues and options for international instruments, below).

Two bodies were instrumental in IPF's work and continue to support the intergovernmental process through IFF: the Interagency Task Force on Forests (ITFF),[35] set up in 1995 to ensure coherent support to IFF by the UN system; and the Forestry Advisers Group (composed of forest advisers of development agencies from donor countries), which helps ensure consistency between international cooperation programmes and IPF proposals.

To structure the support from the UN agencies, ITFF prepared and presented to UNGASS in June 1997 an implementation plan for the IPF proposals addressed to international organizations. This plan, entitled "Interagency partnership on forests: implementation of IPF proposals for action by the ITFF", indicates the agency that will take the lead for each proposal for action (see Box 24).

The Forestry Advisers Group has attempted to structure international cooperation so that it is consistent with IPF's proposals for action. In addition, it has agreed to give high priority to supporting national forest programmes.

The second session of IFF (Geneva, Switzerland, August to September 1998) included review of progress on implementation of IPF proposals; substantive discussions on trade and environment, transfer of technology and forest-related work of

BOX 24

Lead agencies responsible for coordinating follow-up to IPF proposals for action

- ***FAO:*** national forest and land use programmes; fragile ecosystems affected by desertification and drought; impact of airborne pollution on forests; assessment of the multiple benefits of all types of forest; criteria and indicators for sustainable forest management; technology transfer for sustainable forest management (originally under UNDP).
- ***United Nations Environment Programme (UNEP):*** underlying causes of deforestation and land degradation; needs and requirements of countries with low forest cover.
- ***Secretariat of the Convention on Biological Diversity (CBD):*** traditional forest-related knowledge.
- ***United Nations Development Programme (UNDP):*** international cooperation in financial assistance.
- ***Center for International Forestry Research (CIFOR):*** forestry research.
- ***World Bank:*** methodologies for the proper valuation of the multiple benefits of forests.
- ***International Tropical Timber Organization (ITTO):*** trade and environment relating to forest goods and services.

[35] ITFF consists of the following organizations: the Center for International Forestry Research (CIFOR); FAO; the International Tropical Timber Organization (ITTO); the Secretariat of the Convention on Biological Diversity (CBD); the United Nations Department for Social and Economic Affairs; the United Nations Development Programme (UNDP); the United Nations Environment Programme (UNEP); and the World Bank.

international and regional organizations and that being carried out under existing instruments; and background discussions on the programme elements scheduled for substantive discussion at IFF's third session (May 1999). In addition, special attention was paid to the needs of developing countries with low forest cover. Many critical issues, however, remained unresolved, including:

- the modalities of financial cooperation, transfer of technology, the mutually supportive role of trade and environment in the context of sustainable forest management and addressing the underlying causes of deforestation;
- mechanisms for mobilizing the best available information and strengthening research to develop innovative and effective approaches to managing forests as ecological systems;
- the adequacy of present institutions and mechanisms for meeting the complex challenges related to forests.

Various planned government-led initiatives related to these and other issues were announced.

Because of divergent views among countries, a general agreement regarding the need for a global legally binding instrument on forests remained elusive. A government-led initiative on the third category of IFF's programme of work was announced by the governments of Costa Rica and Canada. This initiative will consist of expert meetings and regional consultations on international arrangements and mechanisms to promote the management, conservation and sustainable development of all types of forests. The first global meeting is scheduled to take place in Costa Rica in February 1999, to be followed by regional consultations and a final meeting in Canada in 2000. ◆

Other global and regional initiatives in support of sustainable forest management

DEVELOPMENTS IN INTERNATIONAL CONVENTIONS AND AGREEMENTS[36]

Framework Convention on Climate Change (FCCC)

FCCC, which was adopted at UNCED in 1992, aims at stabilizing the concentration of greenhouse gases in the atmosphere in an effort to prevent human-caused disturbances of the global climatic system. The Convention commits its Parties to carry out national inventories of greenhouse gas emissions and sinks and to work towards meeting their voluntary emission reduction goals. Under FCCC, a pilot phase for "activities implemented jointly" (AIJ) has been established to test and evaluate the feasibility of achieving the Convention's objectives through cooperative projects between Parties, which are designed to avoid, sequester or reduce greenhouse gas emissions. As of 30 August 1998, there were 97 AIJ projects, of which 14 were in the forest sector.[37]

The Kyoto Protocol, adopted at the third Conference of the Parties (Kyoto, December 1997) and awaiting ratification, has taken further steps by establishing legally binding emission reduction commitments for Annex I countries (developed countries and countries with economies in transition). The Kyoto Protocol makes explicit reference to land use change and forestry under several of its articles. (See also the section on environmental and social services of forests in Part I.) The agreements reached give industrialized countries incentives to invest in forestry activities that increase carbon sequestration or reduce carbon emissions from forests. Under the Protocol, a limited list of activities in the land use change and forestry sector can be used to meet the national emissions commitments. Credit towards meeting national commitments can also be given for investment in other countries. Annex I countries may transfer to, or acquire from, other Annex I countries' emission reduction units resulting from projects to reduce greenhouse gas emissions or to enhance greenhouse gas sinks (Article 6). Annex I countries may also invest in emission reduction projects in non-Annex I countries (most developing countries) through the Clean Development Mechanism (CDM) (Article 12), although the extent to which CDM will include land use change and forestry activities is not yet clear.

While many details of how these mechanisms will function remain to be clarified, the possibilities for forestry investment appear promising. Under the Kyoto Protocol, developed countries have continued motivation to reduce emissions from deforestation and to enhance carbon sequestration. Should the decision be taken to include land use change and forestry activities under CDM, carbon offset projects may offer opportunities to developing countries for increased investment in priority forestry activities, technology transfer, reduced petroleum imports, job creation, institutional capacity development and local social and environmental benefits.

By request of the FCCC Subsidiary Body for Scientific and Technical Advice (SBSTA) at its eighth session (June 1998), the Intergovernmental Panel on Climate Change (IPCC) will produce by June 2000 a special report on carbon emissions from sources and removals by sinks from land use, land use change and forestry, to help clarify the implications of the Kyoto Protocol for land use sectors, including the forest sector.

---

[36] The status of ratification of these international conventions and agreements is provided in Table 5 of Annex 3.

[37] These include all projects which have been accepted, approved or endorsed. The forest-sector projects include those listed under the project types "afforestation", "reforestation" and "forest preservation", although some components of energy-sector projects may include forestry-related activities. The source of this information is the FCCC website: http://www.unfccc.de.

Convention on Biological Diversity (CBD)

The fourth Conference of the Parties to CBD was held in Bratislava, Slovakia from 4 to 15 May 1998. Important agreements were reached on international work programmes related to forest biological diversity and marine and coastal area ecosystems, and on the need to complete a protocol on biosafety. Decisions were also taken regarding the assessment of ongoing activities in agrobiodiversity, the importance of the traditional knowledge of indigenous peoples, the equitable sharing of benefits from the use of genetic resources, and the development of a clearing-house mechanism and incentives for the conservation and sustainable use of biological diversity.

The programme of work on forests, the result of a three-year planning effort in three phases, addresses the development of ecosystem approaches to conservation and sustainable use of biodiversity; analysis of the influence of human activities; methodologies needed for elaboration and implementation of criteria and indicators; measures to mitigate biological diversity losses; and ecological landscape models, including protected areas.

Many delegations noted that the proposed programme framework needs further elaboration to become more action oriented. CBD's Subsidiary Body on Scientific, Technical and Technological Advice was requested to report to the sixth Conference of the Parties on the status and trends of forest biological diversity and options for the conservation and sustainable use of forest biological diversity.

United Nations Convention to Combat Desertification (UNCCD)

The first Conference of the Parties to UNCCD was held in Rome from 29 September to 9 October 1997. The conference secured financial resources to support the secretariat, chose the host country and city (Bonn, Germany), decided on the chair and the members of the Committee on Science and Technology, and designated the International Fund for Agricultural Development (IFAD) as the host for the Global Mechanism of UNCCD, a coordinating and implementation-promoting body.

A number of countries, including Burkina Faso, Cape Verde, Chile, China, Mali and Senegal, have begun implementing their National Action Plans. Others (Mali, Tunisia and Uganda) have initiated planning of Convention-related initiatives. The secretariat has a portfolio of countries in Africa, Asia and Latin America in which to promote implementation of the Convention. Several meetings held in preparation for the second Conference of the Parties (Dakar, Senegal, November 1998) have suggested setting up thematic networks to promote information exchange and technical cooperation among developing countries in the areas covered by UNCCD.

Convention on International Trade in Endangered Species of Wild Fauna and Flora (CITES)

Activities connected with CITES's listing of some endangered forest tree species have continued to be discussed in CITES meetings and in a number of other fora. Although several tree species have been listed on CITES appendices for many years (including *Swietenia humilis* and *Swietenia mahagoni*), the subject has become increasingly controversial in recent years with proposals for listing tree species of major commercial importance.

A proposal by Costa Rica to list big-leaf mahogany (*Swietenia macrophylla*) on Appendix III of CITES took effect in 1995. Bolivia and Brazil followed suit; the listing of their populations of *S. macrophylla* on Appendix III became effective in 1998. Appendix III, in practice, implies that the country or countries listing the species must issue export permits and others trading in the species must issue a certificate of origin. As a result, other exporting countries must now issue certificates of origin for their exports of big-leaf mahogany sawlogs, sawnwood and veneers. At the tenth Conference of the Parties to CITES (Harare, Zimbabwe, June 1997), Bolivia and the United States proposed that *S. macrophylla* be listed on Appendix II, which would imply stricter trade restrictions than Appendix III, but the proposal was defeated.

Concern has been expressed that the listing of species under CITES will not, on its own, help save endangered species or populations from extinction. Listing should help promote active protection or management to conserve and enhance the genetic resources under threat of extinction or genetic depletion. Many member countries have stressed that the listing of species should be based on factual and reliable information on threats, distribution, ecology and the genetic variation of species concerned.

These and related concerns were addressed at a Mahogany Working Group meeting held in June 1998 in Brasilia, Brazil as follow-up to the tenth Conference of the Parties. The meeting discussed important topics such as the status of mahogany in the Amazon countries, policies and management practices for mahogany and international cooperation and trade. The meeting recommended a number of actions that would improve cooperation, especially among the Amazon countries.

International Tropical Timber Agreement (ITTA)

The new ITTA 1994, which came into force on 1 January 1997, has been signed by 28 producer countries, 25 consumer countries and the European Union, giving the International Tropical Timber Organization (ITTO) a total membership of 54 (see listing in Table 5 of Annex 3). The agreement remains in force for four years (to 2001) with two three-year extensions possible, which effectively give it a ten-year life.

The current agreement has a greater focus on sustainable forest management than did the previous one (1985 to 1995). The central focus of ITTO's work now is the "Year 2000 Objective", under which all producer countries have made the commitment to have their exports of tropical timber and tropical timber products come from sustainably managed sources by the year 2000. Consumer member countries have also made a commitment to have their forests under sustainable management by 2000. A new fund, the Bali Partnership Fund, has been created to help producing member countries implement sustainable forest management. To date, members have committed about US$12 million (of which US$11.5 million is from Japan); its payment is pending the finalization of the rules for operation.

Among the issues currently being addressed by ITTO are the updating of ITTO's criteria and indicators for sustainable management of natural tropical forests; forest fires; market access; and market difficulties created by the current Asian economic crisis. There is concern that the Asian crisis could hamper producers' efforts to improve forest management practices, thus undermining attempts to achieve ITTO's Year 2000 Objective.

DEVELOPMENTS IN CRITERIA AND INDICATORS FOR SUSTAINABLE FOREST MANAGEMENT

Progress in international processes

Criteria and indicators are tools for assessing national trends in forest conditions and forest management. Whereas "criteria" define the essential components of sustainable forest management, "indicators" are ways to measure or describe a criterion. Together they provide a common framework for describing, monitoring and evaluating progress towards sustainable forest management.

More than 150 countries are currently participating in international processes aimed at the development and implementation of national-level criteria and indicators for sustainable forest management. These efforts are grouped into seven regional and ecoregional initiatives, as follows:

- ITTO, covering the 28 tropical timber producing members of ITTO;
- the Helsinki Process (or Pan-European Process), covering temperate, boreal and Mediterranean forests of 37 European countries (including the Russian Federation) and the European Union;
- the Montreal Process, covering temperate and boreal forests in 12 non-European countries;
- the Tarapoto Proposal, covering tropical forests in the eight member countries of the Amazon Cooperation Treaty;
- the Dry-Zone Africa Process, for dry-zone forests in 28 sub-Saharan countries;
- the Near East Process, for dry-zone forests in 30 Near East countries;
- the Central American Process of Lepaterique, covering all types of forests in the seven member countries of the Central American Commission on Environment and Development (CCAD) and Cuba.

While the initiatives differ somewhat in content

and/or structure,[38] they are similar in objectives and approach. They all incorporate, in some fashion, the following fundamental elements of criteria for sustainable forest management: extent of forest resources; biological diversity; forest health and vitality; productive functions of forests; protective functions of forests; socio-economic benefits and needs; and legal, policy and institutional framework.

The initiatives are in various stages of development and implementation, ranging from initial data collection and reporting on indicators to field testing and consideration of subnational criteria and indicators. Apart from the continuing efforts to refine indicators, the major emphasis of most of the initiatives over the past two years has been the evaluation of the relevance of the criteria and the applicability of the indicators in the light of economic, ecological, social, political and institutional conditions and the needs of the member countries. A number of regional and national meetings and validation workshops have been held for this purpose over the past two years. In addition, the Montreal Process released its first approximation report on implementation in 1997, which examines the relevance of the criteria and indicators to the individual countries.

Other recent developments include a review of the activities related to criteria and indicators for European forests at the Third Ministerial Conference on the Protection of Forests in Europe (Lisbon, Portugal, June 1998). The ministers formally adopted the six Pan-European criteria for sustainable forest management and agreed to develop further the existing Pan-European indicators. The voluntary framework of recommendations for sustainable forest management for practical use (Pan-European Operational Level Guidelines for Sustainable Forest Management) was endorsed, and a commitment was made to the implementation of further activities in support of the Helsinki Process. It is noteworthy that almost all the national reports presented at the Lisbon conference explicitly linked recent changes in their laws and institutions (many passed since 1993) to the Helsinki criteria. The formulation of the criteria for good forest management through the Helsinki Process seems to have resulted in the convergence of European countries regarding management goals and minimum standards.

ITTO, through an Expert Panel established in 1997 by ITTC, has revised its criteria for sustainable tropical forest management in line with recent trends and international developments in the field. ITTC finalized the draft document *Criteria and indicators for the measurement of sustainable management of natural tropical forests* in Libreville, Gabon in May 1998.

Activities at the forest management unit level

National and forest management unit (FMU) level criteria and indicators should be linked and compatible with one another to ensure consistency of approaches for improved forest management. A number of initiatives are under way to identify indicators at the FMU level, including the following.

- The Center for International Forestry Research (CIFOR) is coordinating a project on the identification and testing of indicators for sustainable forest management at the FMU level, with field activities in Brazil, Cameroon, Côte d'Ivoire, India and Indonesia.
- ITTO has financed projects to test sustainable forest management indicators in Brazil, Cameroon, Colombia, Congo, Ecuador, Gabon, Kalimantan (Indonesia), Panama, Peninsular Malaysia, Peru and Sarawak.
- The African Timber Organization (ATO), in an effort to promote sustainable forest management in its member countries, has identified five principles, two subprinciples, 26 criteria and 60 indicators for sustainable forest management at the regional and national levels. It has supported the testing of the indicators at the FMU level in Cameroon, the Congo, Côte d'Ivoire, the Democratic Republic of the Congo, Gabon and Ghana.

The International Union of Forestry Research

[38] Differences among the existing initiatives include the number of national-level criteria (from six in the Helsinki Process to eight in the Central America Process); the level of assessment considered (mostly national, but some initiatives include criteria and indicators for the forest management unit level and/or regional or global level); and the number and array of indicators.

Organizations (IUFRO), in collaboration with FAO and CIFOR, organized an international conference, Fostering Stakeholder Input to Advance Development of Scientifically Based Indicators, in Melbourne, Australia in August 1998. The conference recommended ways of further developing scientifically based indicators for sustainable forest management at the FMU level.

While work has progressed rapidly on conceptualization, elaboration and testing of criteria and indicators for sustainable forest management at the regional and ecoregional levels, with adaption at the national level, additional effort is needed in national- and subnational-level implementation of the criteria and indicators. Opportunities exist for increased cross-initiative support, in which the more experienced processes (i.e. Montreal, Pan-European) could provide additional advice and assistance to the more recent processes.

OTHER INITIATIVES

The G8 countries' action programme on forests

The Heads of State of the G8 countries (Canada, France, Germany, Italy, Japan, the Russian Federation, the United Kingdom and the United States) first proposed the creation of an action programme on forests at their summit in Denver, United States in June 1997. It was subsequently approved at the next summit, held in Birmingham, United Kingdom in May 1998. The action programme commits the G8 countries to take action in five programme areas:

- *Monitoring and assessment of forests:* monitoring and assessment of G8 forests using agreed national criteria and indicators; support of FRA 2000; working with partner countries to build national forest assessment capacity; improving access to remote sensing data and geographic information system technology.
- *National forest programmes:* encouraging partner countries to develop effective national programmes; focusing technical and financial assistance on countries that give priority to sustainable forest management; improving in-country donor coordination.
- *Protected areas:* working to achieve consensus on the purposes and categories of protected forest areas; identification of key forest types not sufficiently represented in protected area systems; strengthening the effectiveness of protected area management.
- *Private sector:* encouraging the private sector to develop voluntary codes of conduct that support sustainable forest management; promoting private investment in sustainable forest management, including innovative financing mechanisms.
- *Illegal logging:* sharing information and assessments on the nature and extent of international trade in illegally harvested timber; implementation of effective measures to combat bribery and corruption in international business transactions.

Model and demonstration forests

Within the framework of the International Model Forest Network, coordinated by Canada, "model forest areas" have been established in several countries, including Canada, Mexico, the Russian Federation and the United States. Argentina, Austria, Chile, China, Ecuador, Indonesia, Japan, Malaysia, the Republic of Karelia in the Russian Federation and Viet Nam are also in the process of establishing model forests. These forests, representing diverse environmental, social and economic conditions, provide an opportunity for comparing concepts and methods and for training in forest management. The network has continued to develop local indicators for assessing the effects of management actions and refining approaches to assessment of the costs and benefits of sustainable forest management.

Another network, which has similar objectives, has been developed by the Tropical Agriculture Research and Higher Education Center (CATIE) in the Central American region under the Demonstration Forest Management Areas initiative.

Eleventh World Forestry Congress

The eleventh World Forestry Congress, hosted by Turkey and organized by Turkey's Ministry of Forestry, was held in Antalya from 13 to 22 October 1997. The World Forestry Congress is held on average every six years, with the support of FAO, and serves as an important technical forum and

BOX 25
The Antalya Declaration

The Antalya Declaration, which was endorsed by the participants of the eleventh World Forestry Congress, called upon countries to demonstrate increased political will to overcome the obstacles to achieving sustainable forest management; upon forestry professionals to respond to the challenges to achieve sustainable forest management; and upon countries and the international donor community to mobilize financial resources and promote environmentally sound and appropriate technology transfer.

The Declaration stressed the importance of raising public awareness of the vital roles of forests to society and of the problems facing them. It supported the implementation of the IPF proposals for action and continued international forest policy dialogue. The Declaration called upon countries to adopt a more cross-sectoral approach to national policies, and to prepare and implement both national forest programmes and national plans for combating desertification. It endorsed improved integration of non-traded forest benefits into markets and the equitable distribution of related costs and benefits. It called for increased attention to the role of women and youth in forestry, and more open and participatory partnerships between all interested parties.

The Declaration called upon research organizations to identify and undertake priority research activities, upon countries to develop and apply national-level criteria and indicators for sustainable forest management and to strengthen national forest inventory and monitoring systems, and upon forest industries to adopt and implement voluntary codes of conduct to contribute to sustainable forest management. It also called for strengthened programmes and supportive policy environments in many technical areas, including community forestry and agroforestry programmes, fast-growing tree plantations, prevention and response to wildfires, rehabilitation of degraded forest land, and conservation and use of biological diversity.

benchmark for developments in the forestry sector. The eleventh congress represented the largest global forestry meeting ever held, attended by over 4 400 participants from 149 countries. The general theme was "Forestry for sustainable development: towards the twenty-first century". An informal Ministerial Meeting, attended by ministers of forestry or their representatives from over 50 countries, and three satellite meetings (conflict management, the role of forestry in combating desertification, and the seventeenth session of Silva Mediterranea) were held before the congress. The results of the congress are recorded in the Antalya Declaration (see Box 25) and in the conclusions and recommendations of the technical sessions.[39]

Silva Mediterranea: 50 years of regional forestry collaboration

Silva Mediterranea, which in 1998 celebrated its fiftieth year as an FAO statutory body, serves as an example of long-term regional collaboration in the forestry sector. As early as 1911 the idea of Mediterranean forestry cooperation was discussed, and in 1922 a Mediterranean forestry league was established under the name "Silva Mediterranea". In 1948, Silva Mediterranea evolved into a statutory body of FAO, officially called the Committee on Mediterranean Forestry Questions. This committee, now consisting of 25 member countries bordering the Mediterranean Sea, as well as the European Union, provides a mechanism for the Mediterranean countries to meet, share their experiences and set cooperative programmes. Among other activities, Silva Mediterranea has drafted a strategic framework – the Mediterranean Forest Action Programme – to assist Mediterranean countries in setting their forest policies. This body not only encourages coordinated efforts in the management of Mediterranean forests, but also helps raise their visibility, promoting awareness of their richness, fragility and susceptibility to threat.

[39] The congress proceedings are available through FAO's website (http://www.fao.org) as well as on CD-ROM.

International NGOs and private-sector initiatives

Private sector. Industry has reacted to the globalization of forestry and forest issues and heightened environmental awareness and concern in a number of ways. It is becoming increasingly involved in initiatives at the national, regional and global levels (e.g. IFF, FCCC) which are likely to have an impact on the forest industry's financial position and/or image. Industry is making significant voluntary efforts to be responsive to the demands of its customers and the public at large for improved management of the forests under its control and for environmentally sound manufacturing technologies. Companies' initiatives are being reported to the public and shareholders through company and industry reports. These voluntary initiatives represent an important contribution to efforts to achieve sustainable forest management and illustrate industry's efforts to be environmentally responsible. The following are three examples of private-sector actions.

Industry-led forest management programmes are appearing in many countries around the world. For example, the American Forest and Paper Association's Sustainable Forestry Initiative has set a recognized standard for the industry across the United States. In Canada, the industry has undertaken a programme aimed at biological diversity conservation through sound forest management and protection of endangered and threatened species. The private sector is also actively supporting sustainable forest management initiatives in various developing countries, as illustrated by some examples in Malaysia. The Malaysian Timber Council (MTC), although not directly responsible for the formulation of certification standards in Malaysia, is contributing to the government's certification efforts. MTC is representing the interests of the Malaysian timber industry and addressing the issues faced by the industry in moving towards sustainable forest management. BP Malaysia has adopted a nation-wide tree planting initiative, in line with the Malaysian national greening campaign. Free seedlings are being distributed with the objective of planting 3 million trees by the year 2000, while a number of parallel initiatives are being contracted out to private companies. In addition, private bodies, such as the Malaysian Business Council for Sustainable Development and the Malaysian International Chamber of Commerce and Industry, sponsor environmental awards to encourage improved environmental performance of Malaysian companies.

ISO 14001, one of the environmental management system standards developed by the International Organization for Standardization (ISO) to assist companies in improving their environmental performance, is rapidly being adopted by industry in various countries (both developed and developing). These standards apply to both forest operations and forest product manufacturing processes. When the ISO 14001 standard is combined with a framework for sustainable forest management, such as criteria and indicators of the Helsinki, Montreal and other regional processes (see preceding discussion in this section), it can become an important globally recognized platform for independently audited forest management certification. This is the conclusion of the ISO Working Group 2 report recently approved by ISO Technical Committee 207. An example of a certification system in line with this thinking is the Canadian Standard Association's system (CSA 808 and CSA 809). Other countries are moving in similar ways (see discussion on certification in Part I).

Industry representatives from around the world are meeting with greater frequency to exchange views and are working together through organizations such as the World Business Council for Sustainable Development (WBCSD), the FAO Advisory Committee on Paper and Wood Products and other less formal fora. Industry representatives recognize the need to portray their concerns in the international forest policy debate in a more concerted way, to the extent that financial considerations allow. There is no doubt, however, that the private sector will continue making progress towards responsible forest management, forest products production and trade.

NGOs. Non-governmental organizations are having an increasing influence on both global and national forest policy by shaping public opinion, engaging in field programmes and lobbying in decision-making fora. Conservation and rural development NGOs now represent the opinions of significant numbers of people in countries throughout the world. The NGOs vary widely, from North-based conservation NGOs with memberships numbering in the tens or hundreds of thousands, to poorly funded grassroots organizations in developing countries working on local issues that often concern human rights as much as forest conservation.

NGOs represent a wide range of opinions, from extremely radical to very conservative. Sometimes quite different NGOs collaborate. There have also been disagreements. For example, organizations interested in human rights and others working on wildlife conservation have differed over issues of indigenous people in protected areas.

Despite this diversity, there are some recognizable trends in approaches taken by NGOs. Since the early 1990s, attention has broadened from the tropics to temperate and boreal forests, and to a more general concern with sustainable forest management. Increasingly, NGOs are consciously trying to balance the needs of human populations with environmental concerns, and many are moving towards a solutions-based approach to forest conservation, rather than grounding their actions in opposition to aspects of current practices. As a result, new partnerships – some which would have seemed highly unlikely a decade ago – have arisen between NGOs and others, such as governments, private corporations and groups of forest dwellers. For example, the World Wide Fund for Nature (WWF) and the World Conservation Union (IUCN) recently spent two years drawing up a detailed picture of forests in the future and a strategy for realizing this vision, published as *Forests for life* (Dudley, Gilmour and Jeanrenaud, 1996). This strategy, which identifies clear targets for progress, has since received the backing of more than 20 governments (with respect to protected area targets), the World Bank and the G8 Summit in Denver, United States in June 1997. The recent signing of an agreement between WWF and the World Bank is a significant example of an alliance between apparently unlikely partners. The development of forest certification and the Forest Stewardship Council is another example in which NGOs have worked with forest industry and forest trade (traditional "enemies"), along with governments, forestry workers and indigenous people, on policies that can benefit all parties concerned.

Not all NGOs support this approach. There is currently considerable debate about the extent to which NGOs should ally themselves with other bodies (e.g. industry through forest certification) or should remain as "outsiders" to promote positive change. While some are seeking greater collaboration, others are intensifying direct action, such as groups blocking logging roads and occupying offices of companies importing endangered timber species. Some of these latter groups reject the more established NGOs. Another question is how much time to devote to international processes, such as CBD and IFF. Should NGOs become a part of the decision-making process or remain as a pressure on this process? The enormous variety of NGOs around the world ensures that different groups will continue to fill different roles.

The World Commission on Forests and Sustainable Development. The World Commission on Forests and Sustainable Development (WCFSD), an independent body of 25 scientists, policy-makers and international specialists from the North and South, was established in 1995. In 1996 and 1997 WCFSD held regional public hearings in Africa, Asia, Europe, Latin America, North America and the Russian Federation to gain a better idea of perceptions about how forests should be managed, for what purposes and for whose benefit. The hearings, which involved representatives from a wide range of interest groups, focused on the political constraints to implementing policy reforms aimed at sustainable forest management and on the interactions among groups that determine forest management outcomes in the political arena. By involving civil society in the debate about global forests, the hearings have provided a useful complement to intergovernmental processes. The work of the commission concluded in 1998. Its report was due to be released in late 1998 or 1999. ♦

Issues and options for international instruments to support sustainable forest management

BACKGROUND

The notion of international forest regulation had its roots in two proposals put forward in 1990: one for a global forest convention, and the other for a forest protocol within the climate change convention. Support for a forest convention increased, particularly among the G7 countries. With this in mind, the Global Legislators Organization for a Balanced Environment (GLOBE) published a model Convention for the Conservation and Wise Use of Forests in early 1992. The hope of various countries that UNCED would produce an international forest convention later that year, however, proved premature; long-running disagreements, especially between developed and developing countries, had become strongly polarized. Countries had – and still have – very different perspectives on what constitute global issues. In part, a country's viewpoint depends on the size of its forest assets and its perceived developmental need to transform some of these assets into other forms of capital. It is also a function of how far the country wishes to treat forests as global commons, as opposed to sovereign domain.

Although a legally binding agreement on forests was not secured, UNCED did produce several legal instruments, both "soft" and "hard" (see Box 26), with direct – if not comprehensive – bearing on the use and management of forests. The Rio Declaration establishes general principles to guide States in a new "equitable global partnership" in matters of environment and development. The Rio principles call for environmental protection to form an integral part of development and for the elimination of unsustainable patterns of production and consumption. Agenda 21 highlights the cross-sectoral nature of forests and underlines the importance of forests in providing socio-economic benefits as well as environmental services. The Forest Principles reaffirm the sovereign right of States to use their forest resources to meet their own needs, while acknowledging that the incremental costs associated with sustainable development must be shared equitably by the international community. The Forest Principles do not, however, identify the global aspects of sustainable forestry. In addition to soft-law instruments, UNCED agreed on the terms of three conventions (CBD, FCCC and UNCCD) which are relevant to, but not solely concerned with, forests and forestry.

These international environmental agreements heralded a period of consensus-building on forests which coalesced around the newly established UN Commission on Sustainable Development. CSD was mandated by ECOSOC to promote the incorporation of the Forest Principles into the implementation of Agenda 21. When CSD established IPF in 1995, it directed IPF to examine future legal instruments as one of its five categories of work. It was not until IPF's final session in February 1997, however, that this mandate was substantively addressed. It became clear then that positions on the subject had shifted somewhat; some countries which previously had been opposed to extending binding forest law were able to consider and even support a legal instrument. Many, however, remained unconvinced. CSD subsequently directed IFF to continue the deliberations on the subject started by IPF, as one of its three categories of work.

BOX 26
Definitions of "hard" and "soft" law

The names of different types of "hard law" tend to reflect negotiating procedure or degree of formality. For example, terms such as "treaty", "convention", "protocol", "agreement", "arrangement" and "statute" are often used interchangeably.

"Convention" refers to a formal legal instrument of a multilateral character. The term also includes instruments adopted by the organs of international institutions such as the International Labour Conference. It is interchangeable with "treaty".

The Vienna Convention of Treaties defines "treaty" as "a formal international agreement concluded between States in written form and governed by international law, whether embodied in a single instrument or in two or more related instruments and whatever its particular designation".

A "protocol" is normally treated as an instrument subsidiary to a convention, or a supplementary treaty concluded at a later date; it is of independent character and operation to the convention and subject to independent ratification.

"Soft law" instruments such as intergovernmental principles (e.g. the Forest Principles) or frameworks for action (e.g. IPF's Proposals for Action) are not legally binding. However, they reflect consensus, negotiated agreement, political endorsement and commitment. Obligations are no more than moral. However, if extensively used, soft law can gain greater legal weight by its cumulative impact on international customary law.

ANALYSIS OF THE ADEQUACY OF LEGAL INSTRUMENTS

Worldwide agreement on what constitutes sustainable forest management at both the national and forest management unit levels has progressed considerably. Various international processes currently working to develop and implement national-level criteria and indicators of sustainable forest management and other efforts to identify indicators for the forest management unit have resulted in considerable convergence of opinion as to what sustainable forest management means in practice. The progress at these two levels has underscored the need to address the global dimensions of sustainable forest management. Failure to ensure sustainable forest management globally will undermine national and subnational initiatives.

A priority at this stage is to address the following questions. What forest issues are best dealt with by international law? Are current instruments adequate? Do they need to be improved? Are new instruments required? If so, what form or forms are most appropriate?

The first step is to distinguish purely global issues from those that are local and national, which is often difficult. Also, the "additionality" component needs to be identified, i.e. what a nation (or company or group) has to do to provide for regional or global needs, beyond what it would do to secure national or local forest goods and services. Reaching agreement on this may be difficult.

The next step is to consider which of the global issues identified can be dealt with only by means of international action. For those issues, the question is whether existing international law is sufficient. If not, there are three broad options:

- "hard law" – taking the form of either forest-specific protocols which extend the obligations defined in related (environmental, developmental and social) conventions, or a stand-alone agreement (i.e. a forest convention) – which would serve primarily to develop and define specific extranational obligations;
- "soft law", such as the Forest Principles and modalities of international cooperation, which might build on the IPF Proposals for Action;
- civil society movements, such as forest certification.

IDENTIFYING INTERNATIONAL FOREST ISSUES

Two central issues that international forest law is well placed to address are: security of global services and other benefits from forests; and cross-border activities and conditions which affect the provision of, or access to, forest goods and services, whether of global, regional, national or local value. These global and cross-border issues can be distinguished from those that are common to many countries but have essentially a local or national nature.

The concept of security of specific forest goods and services is important in clarifying the purpose of future international legal instruments. What people get from forests – and the equitable distribution of associated costs and benefits – matters more than the amount of forest land or deforestation statistics (which have dominated much international discussion).

The security and maintenance of global services from forests are important for humanity as a whole. These services include:

- carbon storage / sequestration for mitigation of global climate change;
- conservation of biological diversity;
- maintenance of the hydrological cycle and soil conservation;
- maintenance of natural heritage and existence value;
- maintenance of forest-related cultural and spiritual values and knowledge.

Although forests fall under domestic jurisdiction, in recent years nations have increasingly acknowledged the "global commons" nature of forests, recognizing the global community's dependence on certain forest functions. International agreements could define national obligations or create incentives for the provision of global forest services, for which new "markets" could develop and which could provide the basis for compensation payments. At present, it is difficult to identify and harness individual countries' willingness to pay for these services. Markets for global services could tip the balance in favour of sustainable forestry, which often provides these services anyway, but which (at present) is not always rewarded for doing so by timber revenues alone. Some markets for forest services, notably for carbon offsets and bioprospecting, have recently developed, offering useful lessons.

A well-established tenet of international law is that nations are responsible for ensuring that activities within their jurisdiction or control do not harm the environment of other nations. Cross-border activities – such as trade, aid, foreign investment, forest operations of foreign companies and pollution – have huge potential to either help or hinder the security of forest goods and services at any level, from local to global. The international legal means to ensure that such activities by one nation (or, indeed, one company) do not create negative impacts in another country are currently insufficient, especially in regard to environmental services of forests. The challenge is to determine what kinds of measures (e.g. agreed standards for operations and procedures, such as environmental impact assessment and codes of practice) or multilateral institutional arrangements might be most effective to address the activities. Cross-border problems have tended to be most effectively addressed through regional agreements in the first instance (e.g. the UN Economic Commission for Europe's Long-Range Transboundary Air Pollution Convention).

PREREQUISITES FOR FURTHER PROGRESS

Five prerequisites will facilitate further progress, regardless of which intergovernmental or civil society options (including the status quo) are selected.

Agreement on criteria and indicators of sustainable forestry

This is necessary so that all relevant international instruments, whatever their locus and form, can be calibrated meaningfully, so that protocols for their communication, payments (for services or as compensation), planning, monitoring and verification can be consistent and can incorporate the critical dimensions of sustainable forestry. Thus, continued support of current efforts to develop criteria and indicators of sustainable forest management at the national and forest management unit level will be important.

Performance and/or gap analysis of international instruments and institutions affecting forests and their links to national and local needs

Any counterproductive policies and procedures should be especially noted.

Development of participatory national forest programmes and international mechanisms to disseminate information on them

Countries need to be clear about what they seek from, and what they can offer to, the international community, while also promoting sustainability and equitability within the nation itself. Strong, stakeholder-driven national forest programmes and associated multistakeholder processes will be fundamental to advancing intergovernmental discussions. Until such national processes are under way and the trade-offs between interest groups are understood, countries will not be well prepared to agree upon analogous trade-offs at the global level. There is a need to ensure that such programmes are not merely wish-lists, but rather that they build upon promising local institutions and mechanisms to reconcile power and potentials for better forestry.

Continued and strengthened engagement in macro- and extrasectoral national policy processes

The sustainability of forests is highly dependent on processes connected to causes of forest problems but emanating from outside the forest sector (e.g. trade and agricultural policy). Engagement with them is essential, whatever mix of international measures is in place.

Involvement in broader trade and development agreements related to international causes of forest problems

Problems linked to trade, aid, foreign investment and pollution have root causes which cannot be tackled solely by international forest agreements. The global issue of indigenous and minority people's rights is a similar case. International forest initiatives clearly need to provide specific inputs on these important issues to bodies such as CSD, the Organisation for Economic Co-operation and Development (OECD) and the World Trade Organization (WTO) and to coordinate fully with them.

OPTIONS FOR AN INTERNATIONAL FOREST REGIME

Four options are laid out here, with no judgement offered on which would be most effective. Analysis and debate by nations as suggested above is critical before any decisions can be made. The first three options cover the status quo or modifications of it. The fourth option touches on the more ambitious option of a global forest convention.

Working through existing conventions and agreements to secure global services and to mitigate harmful cross-border activities

The maintenance and security of some global forest services are provided for by existing global environmental agreements, notably those covering biological diversity conservation (CBD and CITES), wilderness heritage (World Heritage Convention), carbon and climate moderation (FCCC) and protection against desertification (UNCCD). These environmental agreements are increasingly having developmental implications.

CBD provides for forest conservation, notably through setting aside of representative protected areas, but also in the context of sustainable use, rehabilitation and restoration, and equitable access and benefit sharing. As such, the convention is compatible with the requirements of sustainable forest management. Although forest-based biological diversity is only a part of its concerns, the convention has many provisions that can support global services provided by forests, irrespective of whether or not a special forests protocol be established under it. CBD confirms a country's sovereign rights to access and use of biological diversity in its territory, and calls for equitable distribution of the benefits derived from using genetic resources. The convention requires national plans and procedures to integrate biological diversity concerns into sectoral activities. Although it confirms the "global commons" aspect of biological diversity, which paves the way for global payments, adequate incentive mechanisms related to forest use are yet to be established. There are large areas of overlap between the CBD provisions and the Forest Principles, suggesting the possible value of a forest protocol under CBD to solidify obligations for forest conservation and some aspects of sustainable use.

FCCC emphasizes the role of forests as carbon sources and sinks. Through its Joint Implementation projects and the Clean Development Mechanism, the convention is developing implementation and payment protocols covering forest offsets paid for by parties in other countries. Further development of criteria and indicators of sustainable forest management and improved carbon accounting may help widen the options beyond large afforestation or set-aside schemes to include also more complex local forest management and agroforestry schemes as carbon offset projects. Although a protocol for forests under FCCC might be effective for regulating biomass and carbon issues, it would not cover many other forest services.

UNCCD is less focused on specific forest issues, but it emphasizes integrated approaches which address the socio-economic, physical and biological aspects of desertification. By requiring national plans and calling for decentralized resource management, UNCCD has the potential to integrate local and global forest needs. However, it is not applicable to many countries and biomes, and it does not provide comprehensive coverage of all forest services on a global level.

The International Tropical Timber Agreement is a legally binding commodity agreement on tropical forest product trade. As the first commodity agreement to incorporate principles, and now targets, for conserving the resource on which the commodity depends (i.e. tropical forests), ITTA has many elements of a legal agreement on tropical forest management (although reporting mechanisms are not rigorous and there are few real incentives or penalties).

Despite the obvious value of these conventions for forests, relying mainly on this option for an international regime on forests would have limitations. Most of the environmental instruments are relatively new and are not yet fully operational. While several international agreements deal with many global services from forests, they do not cover all of them. In addition, controls on cross-border activities (trade, aid, foreign investment, corporate forestry practices and pollution) are very weak. Finally, there are differences in degree of political commitment, availability of resources for implementation, willingness to pay for the global forest services the instruments provide for, effectiveness and equity of obligations and incentives, and speed and timeliness of actions.

Many parties have been reluctant to let one of the existing conventions (notably CBD and FCCC) take the leading role in an international forest regime because they do not wish forests to be treated primarily as environmental concerns. Such countries (and corporations) – particularly those that value forests highly as economic assets to be maintained, increased or transformed into other forms of capital as best suits the nation – stress the need to consider forests' multiple benefits together so that they can be integrated or traded off according to complex needs. Furthermore, the prospect that parts of the spectrum of benefits and purposes of forests could be divided up and governed by different international instruments and bodies is cause for concern to those who believe that forests have suffered from being treated in a piecemeal fashion.

Relying on "soft law" and civil society processes

Non-legally binding, or "soft law", approaches tend to be employed where room is required to manoeuvre and to experiment. Their main use is as frameworks for action. These approaches are useful in times of rapid market and social change and uncertainty, where local conditions vary greatly and where consensus has not been reached. These conditions characterize many forest issues today, which explains why soft law has been over the past decade, and continues to be, the predominant approach used. Even as soft law, the Forest

Principles and Agenda 21 have increasingly been referred to in conventions since UNCED and have been incorporated into some national laws. Hence they are having an impact on international customary law.

Elaborating, adapting and/or enforcing compliance with intergovernmental soft law, notably the Forest Principles. It has been suggested that the Forest Principles could be adapted and extended to form a basis for an international forest regime. This proposal would appear to have merit for two reasons: unlike other examples of international law, the principles deal exclusively with forests; and they represent a consensus view. However, while the principles focus on forests, they do not deal exclusively (or indeed very much) with global forest issues. Instead, they embody a mix of global, national and local issues, and do not form a basis for regulating global issues. Furthermore, the principles represent a consensus that was achieved precisely because they are not, and were never intended to be, legally binding. Their nature is therefore somewhat incompatible with enforcing compliance, although they are helpful for establishing future agendas and retain the advantages of soft law.

Civil society initiatives. Civil society has already begun to develop incentive systems for better forestry. There are increasing numbers of bilateral markets for biological diversity conservation, bioprospecting and carbon storage (e.g. between corporations and/or NGOs in the industrialized countries and governments in developing countries). An important discussion point is the extent to which these may be alternatives to regulation, and conversely, how much regulation might be needed to ensure their effective and equitable operation.

The forest certification movement has taken off rapidly in some countries through multistakeholder initiatives, albeit driven largely by environmental NGOs, retailers and (more recently) some forestry companies. Certification initiatives could be likened to civil society "conventions" on what sustainable forestry is, how it should be achieved and how managers should be held accountable. While certification has focused on local impacts of commercial forest operations, certification of carbon offset forests has also now begun. Local governments and some companies have also instituted procurement or purchasing policies for forest products, which may or may not include certification. There is certainly scope for these initiatives to give greater attention to global issues. However, there is also a need for national-level frameworks to ensure that these localized initiatives are equitable.

Most civil society initiatives have stressed "best practice" and thereby tend to encourage progressive forest managers. However, NGOs have also called for exposure of bad practice; an example is the Forest Watch proposal to improve transparency regarding corporate logging and forestry activities.

Many of the prevailing problems of international law also apply to civil society initiatives, i.e. different degrees of commitment, resources for implementation, willingness to pay, effectiveness, equity, speed and timeliness. The prerequisites suggested above for further progress – agreement on criteria of sustainable forestry and strong national forest programmes (which can give civil society initiatives a clear legal framework) – must also be in place for these options to be successful. In addition, questions of authority, representativeness, equity and compatibility are important. However, these initiatives are still young, and experimentation and competition among them could be productive for achieving sustainable forest management goals.

The IPF Proposals for Action, representing the product of both nations and civil society, contain the ingredients for establishing ground rules for all soft law activities, whether they be intergovernmental, national or civil.

Developing and/or elaborating regional forest-related legal agreements

Regional agreements have significant potential for addressing cross-border problems, as has been demonstrated in the Central American and Amazon countries' forest agreements and the Pan-European

Process. They can help to create political momentum and institutional conditions that are supportive of long-term investment in sustainable forest management. They also represent bottom-up forms of international cooperation that build on specific shared concerns, i.e. they are very much purpose led and dependent on mutual support. Finally, they tend to make effective and practical links with national forest programmes, which are fundamental to progress in securing forest goods and services at the local, national and global levels. Regional agreements themselves, however, do not address interregional issues and international commitments which are essential for resource transfer between those who benefit from and those who are disadvantaged by forest interventions.

Negotiating a legally binding instrument on all types of forests

A separate forest convention has been proposed as an incentive for States to fulfil their various forest-related commitments under a holistic framework. These obligations would be treated together through a financial and legal mechanism which would allow for progress to be made, reported on and evaluated internationally. Arguments for a forest convention cite the need to deal with forests in their entirety and to avoid fragmentation of forest issues among many initiatives, which can lead to their marginalization.

It seems, however, that it will be difficult in the short term to reach consensus on a broad agenda for an international forest regime. Many issues are genuinely not perceived as priorities by individual nations. In addition, those opposed to a legally binding instrument on forests point out that international agreements are time consuming to negotiate and operate, have opportunity costs and diminishing returns, will weaken other efforts which have been negotiated and implemented at great cost and which need to be consolidated, and/or will promote unsustainable or inequitable objectives.

Forest issues are so diverse and so tied to varied environmental, economic and social interests that hard and soft law, existing and new, will inevitably be mixed. This diversity would call for a flexible framework forest convention which would recognize and support other existing initiatives and bring them together into a holistic picture. Such a convention would not change existing legally binding conventions; rather it would have a gap-filling role by incorporating, step by step, further obligations resulting from continuous international debate and review. The indications to date are that any such framework convention would have to be:

- global in scope;
- non-discriminatory;
- founded on principles of sustainable development;
- focused on the goal of security of goods and services at the global level, and/or on minimizing the international causes of forest problems that affect security of goods and services at any level;
- based on agreed criteria and indicators and involving strong monitoring and learning systems;
- linked to key international fora, but also open to civil society processes;
- complementary to, and harmonized with, other legal agreements;
- backed up by adequate financial and secretariat resources, so that it can be effectively implemented;
- supportive of bottom-up initiatives to develop further and implement the convention.

SOME KEY QUESTIONS FOR CONSIDERATION

The following are some key questions which may be considered in the context of intergovernmental deliberations on international legal instruments on forests.

- What are global forest issues, and how are they distinguished from essentially local or national issues?
- Which global issues lend themselves to international regulation?
- Which global issues are covered by existing international legal instruments?
- Are there civil society alternatives (or complements) to these?

- How effective are existing agreements and civil society instruments, and what potential do they have for improvement?
- What is the feasibility of meaningful consensus on further international regulation?
- What prerequisites need to be in place before an international forest regime can be effective and equitable?

Consideration of these questions will help stimulate analysis and determine whether there is a need for further international law on forests and, if so, what form(s) it might take. ◆

PART IV
FORESTRY IN REGIONAL ECONOMIC GROUPS

This section presents summary information on forest cover and forest products (production and trade) for the 11 regional economic groups listed below. In each case, the figures for imports and exports take into account the value of international trade with all countries, including those within the regional group. The regional economic groups covered are:

- European Union (EU)
- Commonwealth of Independent States (CIS)
- Economic Community of West African States (ECOWAS)
- Common Market for Eastern and Southern Africa (COMESA)
- League of Arab States
- South Asian Association for Regional Cooperation (SAARC)
- Association of Southeast-Asian Nations (ASEAN)
- South Pacific Forum (SPF)
- Latin American Economic System (SELA)
- Caribbean Community and Common Market (CARICOM)
- North American Free Trade Agreement (NAFTA)

European Union (EU)

MEMBERS

Austria, Belgium, Denmark, Finland, France, Germany, Greece, Ireland, Italy, Luxembourg, the Netherlands, Portugal, Spain, Sweden, United Kingdom

DESCRIPTION OF THE FOREST SECTOR

Almost one-third of the region is covered with forest, ranging from boreal to Mediterranean scrub forest. About 70 percent of the forests are located in four countries: Sweden, Finland, France and Germany. Although the EU has only 3 percent of the world's forest area, it is a leading producer of wood-based products, in particular paper and panels. It is a major trader, accounting for 38 percent and 37 percent of the value of world exports and imports of wood-based products, respectively. Sweden, Finland, Germany, France and Austria are among the world's top ten exporters of forest products, and Germany, the United Kingdom, France, the Netherlands, Spain and Belgium/Luxembourg are among the top ten importers.

Forest resources

Land area *(thousand ha)*	313 187
Total forest area, 1995 *(thousand ha)*	102 797
Percent of land in forest *(%)*	32.8
Forest area per caput *(ha)*	0.3
Annual change in forest area, 1990-1995 *(thousand ha)*	340
Percent annual change, 1990-1995 *(%)*	0.3

Production of wood-based products (1996)

Product	Quantity	% of world total
Fuelwood and charcoal	36 360[a]	2
Industrial roundwood	214 726[a]	14
Sawnwood	67 545[a]	16
Wood panels	33 442[a]	22
Pulp for paper	30 661[b]	17
Paper and paperboard	69 826[b]	25

International forest products trade – quantity and value (1996)

Product	Export quantity	Export value		Import quantity	Import value	
		thousand US$	% of world total		thousand US$	% of world total
Fuelwood and charcoal	825[a]	40 716	14	2 389[a]	131 475	47
Industrial roundwood	18 218[a]	1 247 726	11	38 262[a]	2 835 664	21
Sawnwood	28 211[a]	6 661 403	26	30 074[a]	8 043 756	31
Wood panels	12 082[a]	4 809 355	30	12 689[a]	5 296 080	33
Pulp for paper	7 365[b]	3 714 319	23	13 250[b]	7 056 350	38
Paper and paperboard	35 151[b]	35 451 900	55	29 800[b]	27 161 160	44

[a]Thousand m^3.
[b]Thousand tonnes.

Commonwealth of Independent States (CIS)

MEMBERS

Armenia, Azerbaijan, Belarus, Georgia, Kazakhstan, Kyrgyzstan, the Republic of Moldova, the Russian Federation, Tajikistan, Turkmenistan, Ukraine, Uzbekistan

DESCRIPTION OF THE FOREST SECTOR

Nearly one-quarter of the world's forests are located in the CIS countries. The Russian Federation, which alone contains 22 percent of the world's forests, accounts for 94 percent of the forest area in the CIS group. Belarus and Georgia are also heavily forested, but the other countries have low to moderate forest cover. The Russian Federation accounts for about 90 percent of both production and consumption of forest products in the CIS. It is a major exporter of industrial roundwood, accounting for 9 percent of world trade in logs. The other countries are highly dependent on imports to meet their needs for wood products.

Forest resources

Land area *(thousand ha)*	2 178 010
Total forest area, 1995 *(thousand ha)*	809 298
Percent of land in forest *(%)*	37.1
Forest area per caput *(ha)*	2.8
Annual change in forest area, 1990-1995 *(thousand ha)*	501
Percent annual change, 1990-1995 *(%)*	0.1

Production of wood-based products (1996)

Product	Quantity	% of world total
Fuelwood and charcoal	25 101[a]	1
Industrial roundwood	82 695[a]	5
Sawnwood	23 483[a]	5
Wood panels	3 460[a]	2
Pulp for paper	3 771[b]	2
Paper and paperboard	3 355[b]	1

International forest products trade – quantity and value (1996)

Product	Export quantity	Export value		Import quantity	Import value	
		thousand US$	% of world total		thousand US$	% of world total
Fuelwood and charcoal	87[a]	1 524	n.s.	86[a]	3 998	1
Industrial roundwood	16 920[a]	993 022	9	1 003[a]	27 204	n.s.
Sawnwood	5 023[a]	656 242	3	196[a]	27 215	n.s.
Wood panels	1 006[a]	287 376	2	116[a]	38 289	n.s.
Pulp for paper	980[b]	382 839	2	18[b]	14 537	n.s.
Paper and paperboard	1 501[b]	764 485	1	123[b]	118 908	n.s.

[a] Thousand m^3.
[b] Thousand tonnes.

Economic Community of West African States (ECOWAS)

MEMBERS

Benin, Burkina Faso, Cape Verde, the Gambia, Ghana, Guinea, Guinea-Bissau, Côte d'Ivoire, Liberia, Mali, Mauritania, the Niger, Nigeria, Senegal, Sierra Leone, Togo

DESCRIPTION OF THE FOREST SECTOR

The ECOWAS countries together contain only 2 percent of the world's forests. While only 12 percent of the combined land area is in forest, an additional 23 percent is in "other wooded land", which contributes to the production of various forest products, grazing and desertification control. "Trees outside the forest", e.g. in agroforestry systems, are also an important resource. Over 90 percent of the wood produced in ECOWAS countries is used as fuelwood and charcoal. Within ECOWAS, the most important producers of industrial wood products are Nigeria, Côte d'Ivoire and Ghana. ECOWAS accounts for only a minor share of world imports and exports of wood products.

Forest resources

Land area *(thousand ha)*	605 537
Total forest area, 1995 *(thousand ha)*	75 126
Percent of land in forest *(%)*	12.4
Forest area per caput *(ha)*	0.3
Annual change in forest area, 1990-1995 *(thousand ha)*	- 693
Percent annual change, 1990-1995 *(%)*	- 0.9

Production of wood-based products (1996)

Product	Quantity	% of world total
Fuelwood and charcoal	191 711[a]	10
Industrial roundwood	17 360[a]	1
Sawnwood	4 310[a]	1
Wood panels	500[a]	n.s.
Pulp for paper	7[b]	n.s.
Paper and paperboard	57[b]	n.s.

International forest products trade – quantity and value (1996)

Product	Export quantity	Export value		Import quantity	Import value	
		thousand US$	% of world total		thousand US$	% of world total
Fuelwood and charcoal	107[a]	3 701	1	n.s.[a]	14	n.s.
Industrial roundwood	594[a]	95 076	1	32[a]	3 902	n.s.
Sawnwood	771[a]	372 252	1	44[a]	7 806	n.s.
Wood panels	147[a]	82 357	n.s.	42[a]	14 045	n.s.
Pulp for paper	n.s.[b]	254	n.s.	8[b]	4 361	n.s.
Paper and paperboard	n.s.[b]	179	n.s.	143[b]	98 746	n.s.

[a]Thousand m^3.
[b]Thousand tonnes.

Common Market for Eastern and Southern Africa (COMESA)

MEMBERS

Angola, Burundi, the Comoros, the Democratic Republic of the Congo, Djibouti, Egypt, Eritrea, Ethiopia, Kenya, Madagascar, Malawi, Mauritius, Namibia, Rwanda, Seychelles, the Sudan, Swaziland, Uganda, the United Republic of Tanzania, Zambia, Zimbabwe

DESCRIPTION OF THE FOREST SECTOR

COMESA contains 8 percent of the world's forests. While nearly one-fourth of the region's land area is in forest, forest cover varies considerably by country. Countries with high forest cover include the Democratic Republic of the Congo, Uganda, Zambia, the United Republic of Tanzania and Malawi. The others have low to moderate cover. An additional 20 percent of the region's land is in "other wooded land" and some countries have considerable cover of "trees outside the forest", both of which are important sources of wood and non-wood forest products. Over 90 percent of the wood produced in the COMESA region is used as fuelwood and charcoal. COMESA accounts for only a minor share of world imports and exports of wood products.

Forest resources

Land area *(thousand ha)*	1 236 317
Total forest area, 1995 *(thousand ha)*	289 546
Percent of land in forest *(%)*	23.4
Forest area per caput *(ha)*	1.0
Annual change in forest area, 1990-1995 *(thousand ha)*	-2 320
Percent annual change, 1990-1995 *(%)*	- 0.7

Production of wood-based products (1996)

Product	Quantity	% of world total
Fuelwood and charcoal	259 099[a]	14
Industrial roundwood	20 939[a]	1
Sawnwood	1 353[a]	n.s.
Wood panels	315[a]	n.s.
Pulp for paper	440[b]	n.s.
Paper and paperboard	281[b]	n.s.

International forest products trade – quantity and value (1996)

Product	Export quantity	Export value		Import quantity	Import value	
		thousand US$	% of world total		thousand US$	% of world total
Fuelwood and charcoal	8 241[a]	4 477	1	297[a]	2 074	1
Industrial roundwood	210[a]	56 687	n.s.	141[a]	24 043	n.s.
Sawnwood	121[a]	43 094	n.s.	2 645[a]	523 281	2
Wood panels	43[a]	16 221	n.s.	181[a]	68 236	n.s.
Pulp for paper	201[b]	60 098	n.s.	76[b]	32 012	n.s.
Paper and paperboard	7[b]	5 802	n.s.	378[b]	272 371	n.s.

[a] Thousand m^3.
[b] Thousand tonnes.

League of Arab States

MEMBERS

Algeria, Bahrain, the Comoros, Djibouti, Egypt, Iraq, Jordan, Kuwait, Lebanon, the Libyan Arab Jamahiriya, Mauritania, Morocco, Oman, Palestine, Qatar, Saudi Arabia, Somalia, the Sudan, the Syrian Arab Republic, Tunisia, the United Arab Emirates, Yemen

DESCRIPTION OF THE FOREST SECTOR

The combined forest area of these countries accounts for only 1 percent of the world's forests. Some countries (Algeria, Morocco, Somalia, the Sudan, Yemen) have significant areas in "other wooded land", which although not counted in the forest area figures are important for forest products, grazing and desertification control. Fuelwood and charcoal are the major forest products. Production of industrial wood is very limited; about two-thirds of the demand for industrial roundwood and processed wood products is met by imports. Gum arabic is a major export of the Sudan.

Forest resources

Land area *(thousand ha)*	1 351 900
Total forest area, 1995 *(thousand ha)*	50 334
Percent of land in forest *(%)*	3.7
Forest area per caput *(ha)*	0.2
Annual change in forest area, 1990-1995 *(thousand ha)*	-405
Percent annual change, 1990-1995 *(%)*	- 0.8

Production of wood-based products (1996)

Product	Quantity	% of world total
Fuelwood and charcoal	34 739[a]	2
Industrial roundwood	4 274[a]	n.s.
Sawnwood	232[a]	n.s.
Wood panels	347[a]	n.s.
Pulp for paper	450[b]	n.s.
Paper and paperboard	783[b]	n.s.

International forest products trade – quantity and value (1996)

Product	Export quantity	Export value		Import quantity	Import value	
		thousand US$	% of world total		thousand US$	% of world total
Fuelwood and charcoal	244[a]	4 603	1	438[a]	9 280	3
Industrial roundwood	9[a]	1 003	n.s.	690[a]	108 294	1
Sawnwood	14[a]	4 871	n.s.	5 654[a]	1 169 285	4
Wood panels	66[a]	40 041	n.s.	1 556[a]	632 367	4
Pulp for paper	80[b]	45 152	n.s.	433[b]	221 227	1
Paper and paperboard	32[b]	26 053	n.s.	1 696[b]	1 317 967	2

[a] Thousand m^3.
[b] Thousand tonnes.

South Asian Association for Regional Cooperation (SAARC)

MEMBERS

Bangladesh, Bhutan, India, Maldives, Nepal, Pakistan, Sri Lanka

DESCRIPTION OF THE FOREST SECTOR

These countries contain only 2 percent of the world's forest area but support 22 percent of the Earth's population. The region's forest cover is relatively low (19 percent of total land area), but "trees outside the forest", particularly on agricultural land, are an important source of wood and non-wood forest products, notably in Bangladesh, Sri Lanka, Pakistan and parts of India. Over 90 percent of all wood production is used as fuelwood and charcoal, three-quarters of which is consumed in India alone. SAARC accounts for 20 percent of the world production of fuelwood and charcoal. International trade in wood products is limited. India is a major exporter of non-wood forest products, including medicinal plants, essential oils, bidi leaves and lac.

Forest resources

Land area *(thousand ha)*	412 917
Total forest area, 1995 *(thousand ha)*	77 137
Percent of land in forest *(%)*	18.7
Forest area per caput *(ha)*	0.1
Annual change in forest area, 1990-1995 *(thousand ha)*	-141
Percent annual change, 1990-1995 *(%)*	- 0.2

Production of wood-based products (1996)

Product	Quantity	% of world total
Fuelwood and charcoal	370 889[a]	20
Industrial roundwood	29 038[a]	2
Sawnwood	19 453[a]	4
Wood panels	494[a]	n.s.
Pulp for paper	2 173[b]	1
Paper and paperboard	3 559[b]	1

International forest products trade – quantity and value (1996)

Product	Export quantity	Export value		Import quantity	Import value	
		thousand US$	% of world total		thousand US$	% of world total
Fuelwood and charcoal	8[a]	447	n.s.	72[a]	1 582	n.s.
Industrial roundwood	24[a]	2 212	n.s.	477[a]	69 644	n.s.
Sawnwood	28[a]	6 602	n.s.	95[a]	16 698	n.s.
Wood panels	20[a]	10 185	n.s.	54[a]	23 283	n.s.
Pulp for paper	2[b]	1 896	n.s.	360[b]	167 640	1
Paper and paperboard	7[b]	4 839	n.s.	748[b]	445 002	1

[a] Thousand m^3.
[b] Thousand tonnes.

Association of Southeast-Asian Nations (ASEAN)

MEMBERS

Brunei Darussalam, Indonesia, Laos, Malaysia, Myanmar, the Philippines, Singapore, Thailand, Viet Nam

DESCRIPTION OF THE FOREST SECTOR

Nearly half of the ASEAN region is covered with tropical forest, accounting for about 6 percent of the forest area of the world. Three-quarters of the total wood production is used as fuelwood and charcoal. ASEAN is an important timber producing region (particularly of tropical hardwoods), accounting for 6 percent of the world's industrial roundwood production. More than 85 percent of this comes from Indonesia and Malaysia. These two countries are major producers and exporters of wood-based panels (mainly plywood), together accounting for 34 percent (in value) of world export trade in panels. The region is rapidly developing a substantial pulp and paper industry. Many ASEAN countries are major exporters of non-wood forest products, in particular rubber, rattan and bamboo. This region's forests have global significance in terms of biological diversity conservation.

Forest resources

Land area *(thousand ha)*	416 890
Total forest area, 1995 *(thousand ha)*	192 799
Percent of land in forest *(%)*	46.2
Forest area per caput *(ha)*	0.4
Annual change in forest area, 1990-1995 *(thousand ha)*	- 2 748
Percent annual change, 1990-1995 *(%)*	- 1.3

Production of wood-based products (1996)

Product	Quantity	% of world total
Fuelwood and charcoal	294 542[a]	16
Industrial roundwood	97 738[a]	6
Sawnwood	17 854[a]	4
Wood panels	18 418[a]	12
Pulp for paper	3 532[b]	2
Paper and paperboard	8 141[b]	3

International forest products trade – quantity and value (1996)

Product	Export quantity	Export value		Import quantity	Import value	
		thousand US$	% of world total		thousand US$	% of world total
Fuelwood and charcoal	1 822[a]	81 403	28	273[a]	5 008	2
Industrial roundwood	9 526[a]	1 209 375	11	2 034[a]	854 250	6
Sawnwood	5 016[a]	1 868 729	7	4 104[a]	1 117 133	4
Wood panels	14 083[a]	5 733 889	35	1 378[a]	423 490	3
Pulp for paper	1 289[b]	529 847	3	1 177[b]	659 112	4
Paper and paperboard	1 683[b]	1 360 880	2	2 801[b]	2 754 334	4

[a] Thousand m^3.
[b] Thousand tonnes.

South Pacific Forum (SPF)

MEMBERS

Australia, Cook Islands, Fiji, Kiribati, Marshall Islands, Federated States of Micronesia, Nauru, New Zealand, Niue, Palau, Papua New Guinea, Samoa, Solomon Islands, Tonga, Tuvalu, Vanuatu

DESCRIPTION OF THE FOREST SECTOR

The SPF countries together contain less than 3 percent of the world's forest area. Although 86 percent of the region's forest area is located in Australia and Papua New Guinea, several countries are heavily forested – most notably Solomon Islands, Papua New Guinea and Vanuatu, and to a lesser extent Samoa and Fiji. Australia and New Zealand have large plantation programmes. The region (led by Australia, New Zealand and Papua New Guinea) is a significant exporter of industrial roundwood, accounting for 15 percent of world trade. Australia and New Zealand dominate the region's production of industrial roundwood and processed wood products. The small island nations have abundant coconut trees which serve as sources of wood, coconuts, copra and palm oil for local populations.

Forest resources

Land area *(thousand ha)*	846 637
Total forest area, 1995 *(thousand ha)*	89 997
Percent of land in forest *(%)*	10.6
Forest area per caput *(ha)*	3.3
Annual change in forest area, 1990-1995 *(thousand ha)*	- 92
Percent annual change, 1990-1995 *(%)*	- 0.1

Production of wood-based products (1996)

Product	Quantity	% of world total
Fuelwood and charcoal	8 756[a]	n.s.
Industrial roundwood	41 456[a]	3
Sawnwood	6 760[a]	1
Wood panels	2 158[a]	1
Pulp for paper	2 330[b]	1
Paper and paperboard	2 651[b]	1

International forest products trade – quantity and value (1996)

Product	Export quantity	Export value		Import quantity	Import value	
		thousand US$	% of world total		thousand US$	% of world total
Fuelwood and charcoal	12[a]	1 007	n.s.	n.s.[a]	55	n.s.
Industrial roundwood	18 346[a]	1 627 281	15	12[a]	2 775	n.s.
Sawnwood	1 067[a]	303 248	1	808[a]	318 289	1
Wood panels	809[a]	342 270	2	342[a]	113 732	1
Pulp for paper	680[b]	267 333	2	201[b]	152 054	1
Paper and paperboard	625[b]	459 198	1	1 299[b]	1 568 866	2

[a] Thousand m^3.
[b] Thousand tonnes.

Latin American Economic System (SELA)

MEMBERS

Argentina, the Bahamas, Barbados, Belize, Bolivia, Brazil, Chile, Colombia, Costa Rica, Cuba, the Dominican Republic, Ecuador, El Salvador, Grenada, Guatemala, Guyana, Haiti, Honduras, Jamaica, Mexico, Nicaragua, Panama, Paraguay, Peru, Suriname, Trinidad and Tobago, Uruguay, Venezuela

DESCRIPTION OF THE FOREST SECTOR

Forest cover varies greatly within the group, which ranges from very heavily forested countries (Belize, Brazil, Suriname, Guyana) to countries with a small percent of land area in forest (Barbados, Haiti, El Salvador, Uruguay). SELA's membership includes all the Amazon basin countries, which accounts for the region's large amount of forest area, representing 27 percent of the world's forest cover. The countries together are important wood producers, particularly of sawnwood and fuelwood and charcoal (12 and 7 percent respectively of the value of world trade) but also of other wood products. SELA is a significant exporter of paper pulp and fuelwood and charcoal and is a significant importer of paper products. The region's tropical forests have global significance in terms of biological diversity conservation and carbon storage.

Forest resources

Land area *(thousand ha)*	2 006 085
Total forest area, 1995 *(thousand ha)*	941 565
Percent of land in forest *(%)*	46.9
Forest area per caput *(ha)*	2.0
Annual change in forest area, 1990-1995 *(thousand ha)*	-5 802
Percent annual change, 1990-1995 *(%)*	- 0.6

Production of wood-based products (1996)

Product	Quantity	% of world total
Fuelwood and charcoal	255 103[a]	14
Industrial roundwood	142 114[a]	9
Sawnwood	32 783[a]	8
Wood panels	6 948[a]	5
Pulp for paper	10 292[b]	6
Paper and paperboard	12 583[b]	4

International forest products trade – quantity and value (1996)

Product	Export quantity	Export value		Import quantity	Import value	
		thousand US$	% of world total		thousand US$	% of world total
Fuelwood and charcoal	556[a]	20 352	7	78[a]	1 870	1
Industrial roundwood	10 146[a]	542 965	5	200[a]	40 284	n.s.
Sawnwood	4 074[a]	1 089 315	4	1 710[a]	414 070	1
Wood panels	2 253[a]	793 527	5	864[a]	294 407	2
Pulp for paper	4 046[b]	1 953 612	12	1 115[b]	623 682	3
Paper and paperboard	1 855[b]	1 526 489	2	4 267[b]	3 671 159	6

[a]Thousand m^3.
[b]Thousand tonnes.

Caribbean Community and Common Market (CARICOM)

MEMBERS

Antigua and Barbuda, the Bahamas, Barbados, Belize, Dominica, Grenada, Guyana, Jamaica, Montserrat, Saint Lucia, Saint Kitts and Nevis, Saint Vincent and the Grenadines, Suriname, Trinidad and Tobago

DESCRIPTION OF THE FOREST SECTOR

The CARICOM countries account for only 1 percent of the world's forest area, most of which is located in Guyana and Suriname. Belize and Dominica also have a high percentage of their land area in forest. With the exception of Dominica, CARICOM's small island States have moderate to low forest cover. CARICOM's share of world production and trade in forest products is minor. These countries depend heavily on imports to meet their needs for paper, sawnwood and wood-based panels. Guyana is the most active of the countries in terms of forest industries.

Forest resources

Land area *(thousand ha)*	40 504
Total forest area, 1995 *(thousand ha)*	35 843
Percent of land in forest *(%)*	88.5
Forest area per caput *(ha)*	5.8
Annual change in forest area, 1990-1995 *(thousand ha)*	- 51
Percent annual change, 1990-1995 *(%)*	- 0.1

Production of wood-based products (1996)

Product	Quantity	% of world total
Fuelwood and charcoal	500[a]	n.s.
Industrial roundwood	832[a]	n.s.
Sawnwood	183[a]	n.s.
Wood panels	107[a]	n.s.
Pulp for paper	n.s.[b]	n.s.
Paper and paperboard	n.s.[b]	n.s.

International forest products trade – quantity and value (1996)

Product	Export quantity	Export value		Import quantity	Import value	
		thousand US$	% of world total		thousand US$	% of world total
Fuelwood and charcoal	1[a]	17	n.s.	2[a]	94	n.s.
Industrial roundwood	52[a]	8 388	n.s.	47[a]	8 217	n.s.
Sawnwood	27[a]	10 209	n.s.	333[a]	77 642	n.s.
Wood panels	100[a]	37 956	n.s.	89[a]	34 981	n.s.
Pulp for paper	0[b]	0	n.s.	4[b]	3 408	n.s.
Paper and paperboard	1[b]	1 006	n.s.	123[b]	101 001	n.s.

[a] Thousand m^3.
[b] Thousand tonnes.

North American Free Trade Agreement (NAFTA)

MEMBERS

Canada, Mexico, United States

DESCRIPTION OF THE FOREST SECTOR

One-fourth of the area of the NAFTA countries is forest land. Together these three countries account for 15 percent of the world's forest area. Almost 90 percent of the region's forest is located in the United States and Canada, which are major world producers of wood. Canada and the United States together produce 40 percent of the world's industrial roundwood and over one-third of all processed wood products, including almost half the world's paper pulp. NAFTA is extremely important in world trade of wood products. The value of its exports of sawnwood and pulp for paper represents one-half of the world export market. NAFTA accounts for nearly one-third of the world's import market for sawnwood and one-fifth of the paper and paper pulp markets. The forestry sector in these countries is a significant source of revenue and employment through forest industries, forest-based recreation and tourism.

Forest resources

Land area *(thousand ha)*	2 028 878
Total forest area, 1995 *(thousand ha)*	512 473
Percent of land in forest *(%)*	25.3
Forest area per caput *(ha)*	1.3
Annual change in forest area, 1990-1995 *(thousand ha)*	256
Percent annual change, 1990-1995 *(%)*	0.1

Production of wood-based products (1996)

Product	Quantity	% of world total
Fuelwood and charcoal	110 760[a]	6
Industrial roundwood	596 968[a]	40
Sawnwood	175 025[a]	41
Wood panels	50 024[a]	33
Pulp for paper	83 382[b]	46
Paper and paperboard	103 432[b]	37

International forest products trade – quantity and value (1996)

Product	Export quantity	Export value		Import quantity	Import value	
		thousand US$	% of world total		thousand US$	% of world total
Fuelwood and charcoal	682[a]	38 411	13	396[a]	20 027	7
Industrial roundwood	23 677[a]	2 884 452	26	9 315[a]	472 598	3
Sawnwood	58 024[a]	11 862 320	47	46 211[a]	7 773 383	30
Wood panels	9 715[a]	2 386 590	15	9 614[a]	2 398 628	15
Pulp for paper	15 706[b]	8 213 371	51	5 741[b]	3 850 166	19
Paper and paperboard	22 973[b]	16 945 310	26	14 368[b]	11 902 824	21

[a] Thousand m^3.
[b] Thousand tonnes.

ANNEXES

Annex 1

ACRONYMS

ANZCERTA
Australia-New Zealand Closer Economic Relations Trade Agreement
ASEAN
Association of Southeast-Asian Nations
ATO
African Timber Organization
CARICOM
Caribbean Community and Common Market
CATIE
Tropical Agriculture Research and Higher Education Center
CBD
Convention on Biological Diversity
CCAD
Central American Commission on Environment and Development
CDM
Clean Development Mechanism
CGIAR
Consultative Group on International Agricultural Research
CIFOR
Center for International Forestry Research
CIS
Commonwealth of Independent States
CITES
Convention on International Trade in Endangered Species of Wild Fauna and Flora
COFO
FAO's Committee on Forestry
COMESA
Common Market for Eastern and Southern Africa
CSD
Commission on Sustainable Development
ECE
UN Economic Commission for Europe
ECOSOC
United Nations Economic and Social Council
ECOWAS
Economic Community of West African States
EU
European Union
FAG
Forestry Advisers Group
FCCC
Framework Convention on Climate Change
FORSPA
Forestry Research Support Programme for Asia and the Pacific
FRA
Forest Resources Assessment
GATT
General Agreement on Tariffs and Trade
GDP
gross domestic product
GFAR
Global Forum on Agricultural Research
GFIS
Global Forest Information System
GFSM
Global Fibre Supply Model
GNP
gross national product
IARC
international agricultural research centre
ICDP
Integrated Conservation Development Project
ICRAF
International Centre for Research in Agroforestry
IFF
Intergovernmental Forum on Forests
ILO
International Labour Organisation
IPCC
Intergovernmental Panel on Climate Change
IPF
Intergovernmental Panel on Forests
IPGRI
International Plant Genetic Resources Institute
ISO
International Organization for Standardization
ITFF
Inter-Agency Task Force on Forests

ITTA
International Tropical Timber Agreement
ITTC
International Tropical Timber Council
ITTO
International Tropical Timber Organization
IUCN
International Union for Conservation of Nature and Natural Resources (World Conservation Union)
IUFRO
International Union of Forestry Research Organizations
NAFTA
North American Free Trade Agreement
NARS
national agricultural research systems
NFP
national forest programme
NGO
non-governmental organization
NWFP
non-wood forest product
OECD
Organisation for Economic Co-operation and Development
SAARC
South Asian Association for Regional Cooperation
SELA
Latin American Economic System
SPARTECA
South Pacific Regional Trade and Economic Cooperation Agreement
SPF
South Pacific Forum
UNCCD
Convention to Combat Desertification
UNCED
United Nations Conference on Environment and Development
UNCTAD
United Nations Conference on Trade and Development
UNDP
United Nations Development Programme
UNEP
United Nations Environment Programme
UNGASS
Special Session of the United Nations General Assembly
WCFSD
World Commission on Forests and Sustainable Development
WCMC
World Conservation Monitoring Centre
WRI
World Resources Institute
WTO
World Trade Organization
WWF
World Wide Fund for Nature

Annex 2

DEFINITIONS

COUNTRY GROUPS

Reference to regions and subregions correspond to the country groups indicated in Tables 1 and 2 of Annex 3. These in turn use the definitions of the FAOSTAT country groups list of FAO.

Developed countries

Includes all industrialized countries and countries in transition.

Industrialized countries

Includes all countries in Europe except Eastern Europe (see country listing in Table 2 of Annex 3) plus Australia, Canada, Israel, Japan, New Zealand, South Africa and the United States.

Countries in transition

Includes all countries in Eastern Europe (see country listing in Table 2 of Annex 3) and Armenia, Azerbaijan, Georgia, Kazakhstan, Kyrgyzstan, Tajikistan, Turkmenistan and Uzbekistan.

Developing countries

Includes all countries other than developed countries.

Tropical countries

As indicated in Table 2 of Annex 3.

Temperate/non-tropical countries

As indicated in Table 2 of Annex 3.

FOREST AND RELATED LAND COVER DEFINITIONS

These definitions have been adopted for use in FAO's Global Forest Resources Assessment (FRA). They apply to FRA 1990 and the 1995 set of forest cover data provided in Tables 2 and 3 of Annex 3. Some terms have two definitions, one when used in conjunction with developing countries and another when applied to developed countries, because of the different data collection methods used by the two organizations involved: FAO (for developing countries) and the UN Economic Commission for Europe (ECE) Trade Division Timber Section (for developed countries). FRA 2000 is using a common set of definitions, some of which differ slightly from those given below. The new set of definitions is available in the document *UN-ECE/FAO Temperate and Boreal Forest Resources Assessment 2000: terms and definitions* (UN, 1997b).

Afforestation/reafforestation

The establishment of a tree crop on an area from which it has always or very long been absent. Where such establishment fails and is repeated, the latter may properly be termed reafforestation.

Reforestation

Establishment of a tree crop on forest land.

FOREST AND RELATED LAND COVER DEFINITIONS – DEFINITIONS APPLIED FOR DEVELOPED COUNTRIES

Forest

Includes natural forest and forest plantations. Land with tree crown cover (stand density) of more than about 20 percent of the area; continuous forest with trees usually growing to more than about 7 m in height and able to produce wood. This includes both closed forest formations where trees of various storeys and undergrowth cover a high proportion of the ground, and open forest formations with a continuous grass layer in which tree synusia cover at least 10 percent of the ground.

Deforestation

Change of forest with depletion of tree crown cover to less than 20 percent.

Other wooded land

Land which has some forestry characteristics but is not forest as defined above. It includes open woodland and scrub, shrub and brushland (see below), whether or not used for pasture or range. It excludes land occupied by "trees outside the forest".

Open woodland

Land with tree crown cover (stand density of about 5 to 20 percent of the area).

Scrub, shrub and brushland

Land with scrub, shrub or stunted trees, where the main woody elements are shrubs (usually more than 50 cm and less than 7 m in height), covering more than about 20 percent of the area, not primarily used for agricultural or other non-forestry purposes, such as grazing of domestic animals.

Exploitable forest

Forest and other wooded land on which there are no legal, economic or technical restrictions on wood production. It includes areas where, although there are no such restrictions, harvesting is not currently taking place, for example, areas included in long-term utilization plans or intentions.

FOREST AND RELATED LAND COVER DEFINITIONS – DEFINITIONS APPLIED FOR DEVELOPING COUNTRIES

Forest

Ecosystem with a minimum of 10 percent crown cover of trees and/or bamboos, generally associated with wild flora and fauna and natural soil conditions, and not subject to agricultural practices. The term forest is further subdivided, according to its origin, into two categories:

- Natural forests: a subset of forests composed of tree species known to be indigenous to the area.
- Plantation forests: established artificially by afforestation on lands which previously did not carry forest within living memory, or established artificially by reforestation of land which carried forest before, with replacement of the indigenous species by a new and essentially different species or genetic variety.

Deforestation

Change of forest with depletion of tree crown cover to less than 10 percent. (Changes within the forest class, e.g. from closed to open forest, which negatively affect the stand or site and, in particular, lower the production capacity, are termed forest degradation and are considered apart from deforestation.)

Other wooded land

Includes the following:

- Forest fallow, consisting of all complexes of woody vegetation deriving from the clearing of natural forest for shifting agriculture. It consists of a mosaic of various succession phases and includes patches of uncleared forests and agricultural fields which cannot be realistically segregated and accounted for area wide, especially from satellite imagery. Forest fallow is an intermediate class between forest and non-forest land uses. Part of the area which is not under cultivation may have the appearance of a secondary forest. Even the part currently under cultivation sometimes has the appearance of forest, because of the presence of tree cover. Accurate separation between forest and forest fallow may not always be possible.
- Shrubs, referring to vegetation types where the dominant woody elements are shrubs with more than 50 cm and less than 5 m height on maturity. The height limits for trees and shrubs should be interpreted with flexibility, particularly where the minimum tree and maximum shrub heights may vary between 5 and 7 m approximately.

FOREST PRODUCTS DEFINITIONS

These definitions are an abbreviated form of those used by the *FAO Yearbook of Forest Products*. See a recent volume of that publication for full details (e.g. FAO, 1998g).

Roundwood

Wood in its natural state as removed from forests and from trees outside the forest; wood in the rough. Commodities include all forms of industrial roundwood and fuelwood.

Fuelwood and charcoal

Includes "wood in the rough" (from trunks and branches of trees) to be used as fuel such as cooking, heating or power production. Wood for charcoal is included.

Woodfuel

An aggregate term including fuelwood, charcoal and black liquor.

Black liquor

The alkaline-spent liquor obtained from the digesters in the production of sulphate or soda pulp during the process of paper production.

Industrial roundwood

The commodities included are sawlogs or veneer logs, pulpwood and other industrial roundwood. In the case of trade, chips and particles and wood residues are also included.

Sawnwood

Wood (including sleepers) sawn lengthwise or produced by a profile-chipping process, and planed wood.

Wood-based panels

An aggregate term including the following commodities: veneer sheets, plywood, particle board and fibreboard. Particleboard includes varieties such as oriented strand board and flakeboard. Fibreboard includes hardboard, medium-density fibreboard and insulation fibreboard.

Pulp for paper

Includes both wood pulp (mechanical, semi-chemical and chemical) and other fibre pulp (straw, bamboo and bagasse, etc.).

Recovered paper

Used paper and paperboard or residues from paper conversion, collected for reuse as a raw material for the manufacture of paper, paperboard or other products.

Paper and paperboard

The following commodities are included in this aggregate: newsprint, printing and writing paper, other paper and paperboard.

SPECIES DEFINITIONS

Coniferous species (softwoods)

All woods derived from trees classified botanically as Gymnospermae. These are generally referred to as softwoods.

Non-coniferous species (hardwoods)

All woods derived from trees classified botanically as Angiospermae. These are generally referred to as broadleaved or hardwoods.

OTHER DEFINITIONS

FAO has no formally adopted definitions for the following terms. These definitions are provided to clarify their meaning in the context of the *State of the World's Forests 1999*.

Civil society

Non-State organizations and movements, including citizen groups, NGOs, citizens' organizations (formal and informal), mass media, business leaders, the research community and social and political movements.

Pluralism

The existence within any society of a variety of groups with different, autonomous and sometimes conflicting interests, values and perspectives. These differing views cannot be reduced to a common perspective by the reference to an absolute standard.

Stakeholder

An individual, social group or institution that possesses a direct, significant and specific interest in a resource or a service.

Annex 3

DATA TABLES

GENERAL NOTES

Country nomenclature and regional groups used in the tables

The country names and order used in these tables follow standard UN practice regarding nomenclature and alphabetical listing of countries. Separate data are not given for "Taiwan Province of China", but are incorporated in the data for China, as consistent with UN practice. The regional groups used in these tables represent FAO's standardized regional breakdown of the world according to geographical (note: not economic or political) criteria. Tables 2 and 3 further break down these regions into subregional groups. This breakdown parallels ecofloristic zones and is used to provide the user with additional information about forest resources within a given region.

Totals

Numbers may not tally because of rounding.

Abbreviations

n.s. = not significant, indicating a very small value
n.a. = not available
n.ap. = not applicable

TABLE 1: BASIC COUNTRY DATA

The table gives data for all regions: Africa, Asia, Oceania, Europe, North and Central America, and South America. The land area figure refers to total area of the country, excluding areas under inland water bodies. The source of these data is the *FAO Production Yearbook 1996* (FAO, 1998h). Population statistics on total population, population density and annual rate of change are taken from *World population prospects – the 1996 revision* (UN, 1996). The source of percent rural population data is the *World urbanization prospects – the 1996 revision* (UN, 1997a).

The source of the economic data is *World development indicators 1997* (World Bank, 1997). The gross national product (GNP) per caput figure represents the GNP converted to United States dollars using the World Bank Atlas method, divided by the mid-year population. The annual growth rate figures for gross domestic product (GDP) are calculated using constant price data in local currency.

TABLE 2: FOREST COVER, 1995

The table gives data for all regions, broken down by country and by ecofloristic zone. These 1995 figures represent the most current global data set on forest cover and forest cover change available. The source of the data is the FAO Forest Resources Assessment Project, which published them in the *State of the World's Forests 1997* (FAO, 1997d). The methods used for collecting these data are described in that publication.

The Pacific Islands Trust Territory data in this table represent the aggregate figures for Northern Mariana Islands, Micronesia, the Marshall Islands and Palau. Although the Pacific Island Trust Territories entity was dissolved in 1994, disaggregate figures for the four components were not available at the time the table was originally prepared. "Total forest" is the sum of natural forest and plantations. Because of difficulties in drawing a clear distinction between the two in many developed countries, only "total forest" figures are provided for developed countries and "n.ap." (not applicable) is indicated under "natural forest". The "world total" figure for land area is greater than the sum of the land areas of the listed countries, because not all countries of the world are included in the table.

The subregional breakdown reflects ecofloristic zones, not economic or political groupings.

TABLE 3: CHANGE IN FOREST COVER, 1990-1995

The table gives data for all regions, broken down by country and by ecofloristic zone. See notes to Table 2.

TABLE 4: PRODUCTION, TRADE AND CONSUMPTION OF FOREST PRODUCTS, 1996

The table gives data for all regions, broken down by country. The source of these data is the *FAO Yearbook of Forest Products 1996* (FAO, 1998g). The information is also available electronically through FAO's website: http://www.fao.org.

TABLE 5: INTERNATIONAL CONVENTIONS AND AGREEMENTS AND NATIONAL FOREST PROGRAMMES

The table gives countries that have ratified the UNCED conventions – the Convention on Biological Diversity (CBD), the Framework Convention on Climate Change (FCCC) and the Convention to Combat Desertification (UNCCD); countries that are members under the International Tropical Timber Agreement (ITTA); and countries that have national forest programmes (NFPs).

The reference date for ratification status and the source of information on the conventions and agreements are as follows:

- CBD: status as of 28 May 1998. The source of this information is the CBD website: http://www.biodiv.org.
- FCCC: status as of 20 July 1998. The source of this information is the FCCC website: http://www.unfccc.de.
- UNCCD: status as of 31 August 1998. The source of this information is the UNCCD website: http://www.unccd.ch/lite.
- ITTA 1994: status as of 7 September 1998. The source of this information is ITTO, personal communication.

In addition to the countries indicated, the European Community has ratified CBD, FCCC and UNCCD and the European Union has signed ITTA 1994.

The regional breakdown reflects geographic rather than economic or political groupings.

Abbreviations used in Table 5

FSR = Forestry Sector Review
MP = Forestry Sector Master Plan
NCS = National Conservation Strategy
NEAP = National Environmental Action Plan
NEMS = National Environmental Management Strategy
NFAP = National Forestry Action Programme (including Tropical Forests Action Programme, TFAP)
NP = National Legal Policy or Planning Framework
NPCD = National Plan to Combat Desertification
* = Planning process interrupted, not completed

TABLE 1

Basic country data

Country	Land area	Population				Economic indicators	
	Total 1996 *(thousand ha)*	Total 1997 *(million)*	Density 1997 *(population/ km²)*	Annual rate of change 1995-2000 *(%)*	Rural 1997 *(%)*	GNP per caput 1995 *(US$)*	Annual growth rate of GDP 1990-1995 *(%)*
AFRICA							
Algeria	238 174	29.5	12.4	2.3	42.9	1 600	0.1
Angola	124 670	11.6	9.3	3.3	67.8	410	-4.1
Benin	11 062	5.7	51.5	2.8	60.1	370	4.1
Botswana	56 673	1.5	2.6	2.2	34.9	3 020	4.2
Burkina Faso	27 360	11.0	40.2	2.8	83.1	230	2.6
Burundi	2 568	6.4	249.2	2.8	91.9	160	-2.3
Cameroon	46 540	13.9	29.9	2.7	53.6	650	-1.8
Cape Verde	403	0.4	99.3	2.5	42.9	960	n.a.
Central African Republic	62 298	3.4	5.5	2.1	60.1	340	1.0
Chad	125 920	6.7	5.3	2.8	77.2	180	1.9
Comoros	223	0.6	269.1	3.1	68.5	470	n.a.
Congo	34 150	2.7	7.9	2.8	40.0	680	-0.6
Congo, Democratic Republic of	226 705	48.0	21.2	2.6	70.7	120	n.a.
Côte d'Ivoire	31 800	14.3	45.0	2.0	55.3	660	0.7
Djibouti	2 318	0.6	25.9	2.7	17.5	n.a.	n.a.
Egypt	99 545	64.5	64.8	1.9	54.9	790	1.3
Equatorial Guinea	2 805	0.4	14.3	2.5	55.5	380	n.a.
Eritrea	10 100	3.4	33.7	3.7	82.3	n.a.	n.a.
Ethiopia	100 000	60.1	60.1	3.2	83.7	100	n.a.
Gabon	25 767	1.1	4.3	2.8	47.9	3 490	-2.5
Gambia	1 000	1.2	120.0	2.3	69.6	320	1.6
Ghana	22 754	18.3	80.4	2.8	63.1	390	4.3
Guinea	24 572	7.6	30.9	1.3	69.4	550	3.8
Guinea-Bissau	2 812	1.1	39.1	2.0	77.5	250	3.5
Kenya	56 914	28.4	49.9	2.2	69.7	280	1.4
Lesotho	3 035	2.1	69.2	2.5	74.5	770	7.5
Liberia	9 632	2.5	26.0	8.6	53.9	n.a.	n.a.
Libyan Arab Jamahiriya	175 954	5.8	3.3	3.3	13.7	n.a.	n.a.
Madagascar	58 154	15.8	27.2	3.1	72.4	230	0.1
Malawi	9 408	10.1	107.4	2.5	85.8	170	0.7
Mali	122 019	11.5	9.4	3.0	71.9	250	2.5
Mauritania	102 522	2.4	2.3	2.5	46.3	460	4.0
Mauritius	203	1.1	541.9	1.1	59.2	3 380	4.9
Morocco	44 630	27.5	61.6	1.8	46.8	1 110	1.2
Mozambique	78 409	18.3	23.3	2.5	63.7	80	7.1
Namibia	82 329	1.6	1.9	2.4	62.1	2 000	3.8
Niger	126 670	9.8	7.7	3.3	80.9	220	0.5
Nigeria	91 077	118.4	130.0	2.8	58.7	260	1.6
Réunion	250	0.7	280.0	1.3	31.1	n.a.	n.a.
Rwanda	2 467	5.9	239.2	7.8	94.1	180	-12.8
Saint Helena	31	n.s.	n.s.	n.a.	33.3	n.a.	n.a.
Sao Tome and Principe	96	0.1	104.2	n.a.	55.8	350	n.a.
Senegal	19 253	8.8	45.7	2.7	55.0	600	1.9
Seychelles	45	0.1	222.2	n.a.	44.6	6 620	n.a.
Sierra Leone	7 162	4.4	61.4	3.0	65.4	180	-4.2
Somalia	62 734	10.2	16.3	3.9	73.6	n.a.	n.a.
South Africa	122 104	43.3	35.5	2.2	50.3	3 160	0.6
Sudan	237 600	27.9	11.7	2.2	66.8	n.a.	6.8
Swaziland	1 720	0.9	52.3	2.8	67.2	1 170	n.a.
Togo	5 439	4.3	79.1	2.7	68.3	310	-3.4

Note: The regional breakdown reflects geographical rather than economic or political groupings.

Country	Land area	Population				Economic indicators	
	Total 1996 *(thousand ha)*	Total 1997 *(million)*	Density 1997 *(population/ km²)*	Annual rate of change 1995-2000 *(%)*	Rural 1997 *(%)*	GNP per caput 1995 *(US$)*	Annual growth rate of GDP 1990-1995 *(%)*
Tunisia	15 536	9.3	59.9	1.8	36.7	1 820	3.9
Uganda	19 965	20.8	104.2	2.6	86.8	240	6.6
United Republic of Tanzania	88 359	31.5	35.7	2.3	74.4	120	3.2
Zambia	74 339	8.5	11.4	2.4	56.4	400	-0.2
Zimbabwe	38 685	11.7	30.2	2.1	66.9	540	1.0
ASIA							
Afghanistan	65 209	22.1	33.9	5.3	79.3	670	n.a.
Armenia	2 820	3.6	127.7	0.2	30.9	730	-21.2
Azerbaijan	8 660	7.6	87.8	0.8	43.7	480	-20.2
Bahrain	69	0.6	869.6	2.1	8.9	7 840	n.a.
Bangladesh	13 017	122.0	937.2	1.6	80.6	240	4.1
Bhutan	4 700	1.9	40.4	2.8	93.5	420	n.a.
Brunei Darussalam	527	0.3	56.9	2.1	29.6	25 160	n.a.
Cambodia	17 652	10.5	59.5	2.2	78.4	270	6.4
China	932 641	1 243.7	133.4	0.9	68.2	620	12.8
China – Hong Kong SAR	99	6.2	6 262.6	0.8	4.7	22 990	5.6
Cyprus	924	0.8	86.6	1.2	44.9	n.a.	n.a.
DPR of Korea	12 041	22.8	189.4	1.6	38.2	n.a.	n.a.
Georgia	6 970	5.4	77.5	-0.1	40.7	440	-26.9
India	297 319	960.2	323.0	1.6	72.6	340	4.6
Indonesia	181 157	203.5	112.3	1.5	62.7	980	7.6
Iran, Islamic Republic of	162 200	71.5	44.1	2.2	40.0	n.a.	4.2
Iraq	43 737	21.2	48.5	2.8	24.6	n.a.	n.a.
Israel	2 062	5.8	281.3	1.9	9.1	15 920	6.4
Japan	37 652	125.6	333.6	0.2	21.6	39 640	1.0
Jordan	8 893	4.5	50.6	3.3	27.5	1 510	8.2
Kazakhstan	267 073	16.8	6.3	0.1	39.6	1 330	-11.9
Kuwait	1 782	1.7	95.4	3.0	2.8	17 390	12.2
Kyrgyzstan	19 180	4.5	23.5	0.4	60.7	700	-14.7
Laos	23 080	5.2	22.5	3.1	78.2	350	6.5
Lebanon	1 023	3.1	303.0	1.8	11.6	2 660	n.a.
Macau	2	0.4	20 000.0	2.1	1.3	n.a.	n.a.
Malaysia	32 855	21.0	63.9	2.0	45.0	3 890	8.7
Maldives	30	0.3	1 000.0	3.4	72.5	990	n.a.
Mongolia	156 650	2.6	1.7	2.1	38.1	310	-3.3
Myanmar	65 755	46.8	71.2	1.8	73.4	n.a.	5.7
Nepal	14 300	22.6	158.0	2.5	89.1	200	5.1
Oman	21 246	2.4	11.3	4.2	21.2	4 820	6.0
Pakistan	77 088	143.8	186.5	2.7	64.6	460	4.6
Philippines	29 817	70.7	237.1	2.0	44.2	1 050	2.3
Qatar	1 100	0.6	54.5	1.8	8.1	11 600	n.a
Republic of Korea	9 873	45.7	462.9	0.9	16.8	9 700	7.2
Saudi Arabia	214 969	19.5	9.1	3.4	16.1	7 040	1.7
Singapore	61	3.4	5 573.8	1.5	0	26 730	8.7
Sri Lanka	6 463	18.2	281.6	1.0	77.3	700	4.8
Syrian Arab Republic	18 378	14.9	81.1	2.5	46.9	1 120	7.4
Tajikistan	14 060	6.0	42.7	1.9	67.5	340	-18.1
Thailand	51 089	59.1	115.7	0.8	79.4	2 740	8.4
Turkey	76 963	62.8	81.6	1.5	28.4	2 780	3.2
Turkmenistan	46 993	4.2	8.9	1.9	54.9	920	-10.6
United Arab Emirates	8 360	2.3	27.5	2.0	15.3	17 400	n.a.
Uzbekistan	41 424	23.6	57.0	1.9	58.4	970	-4.4
Viet Nam	32 549	76.5	235.0	1.8	80.4	240	8.3
Yemen	52 797	18.3	34.7	3.7	64.8	260	n.a.

Note: The regional breakdown reflects geographical rather than economic or political groupings.

Country	Land area	Population				Economic indicators	
	Total 1996 *(thousand ha)*	Total 1997 *(million)*	Density 1997 *(population/ km²)*	Annual rate of change 1995-2000 *(%)*	Rural 1997 *(%)*	GNP per caput 1995 *(US$)*	Annual growth rate of GDP 1990-1995 *(%)*
OCEANIA							
American Samoa	20	0	0	n.a.	49.1	n.a.	n.a.
Australia	768 230	18.2	2.4	1.1	15.3	18 720	3.5
Cook Islands	23	0	0	n.a.	36.8	n.a.	n.a.
Fiji	1 827	0.8	43.8	1.6	58.7	2 440	n.a.
French Polynesia	366	0.2	54.6	1.9	43.6	n.a.	n.a.
Guam	55	0.1	181.8	1.8	60.9	n.a.	n.a.
Kiribati	73	0.1	137.0	n.a.	64.2	920	n.a.
New Caledonia	1 828	0.2	10.9	1.5	37.1	n.a.	n.a.
New Zealand	26 799	3.6	13.4	1.1	13.7	14 340	3.6
Niue	26	n.s.	n.s.	n.a.	50.0	n.a.	n.a.
Pacific Islands (Trust Terr.)	178	n.s.	n.s.	n.a.	n.a.	n.a.	n.a.
Papua New Guinea	45 286	4.5	9.9	2.2	83.4	1 160	9.3
Samoa	283	0.2	70.7	1.1	78.6	n.a.	n.a.
Solomon Islands	2 799	0.4	14.3	3.2	81.9	910	n.a.
Tonga	72	0.1	138.9	n.a.	56.6	1 630	n.a.
Vanuatu	1 219	0.2	16.4	2.5	80.9	1 200	n.a.
EUROPE							
Albania	2 740	3.4	124.1	0.6	62.0	670	1.4
Austria	8 273	8.2	99.1	0.6	35.5	26 890	1.9
Belarus	20 748	10.3	49.6	-0.1	27.6	2 070	-9.3
Belgium and Luxembourg	3 282	10.6	323.0	0.3	3.1	24 710	1.1
Bosnia and Herzegovina	5 100	3.8	74.5	3.9	58.1	n.a.	n.a.
Bulgaria	11 055	8.4	76.0	-0.5	31.0	1 330	-4.3
Croatia	5 592	4.5	80.5	-0.1	43.4	3 250	n.a.
Czech Republic	7 728	10.2	132.0	-0.1	34.2	3 870	-2.6
Denmark	4 243	5.2	122.6	0.2	14.6	29 890	2.0
Estonia	4 227	1.4	33.1	-1.0	26.5	2 860	-9.2
Finland	30 459	5.1	16.7	0.3	36.1	20 580	-0.5
France	55 010	58.5	106.3	0.3	25.0	24 990	1.0
Germany	34 927	82.2	235.3	0.3	13.1	27 510	n.a.
Greece	12 890	10.5	81.5	0.3	40.4	8 210	1.1
Hungary	9 234	10.0	108.3	-0.6	34.5	4 120	-1.0
Iceland	10 025	0.3	3.0	1.0	8.4	24 950	n.a.
Ireland	6 889	3.5	50.8	0.2	42.1	14 710	4.7
Italy	29 406	57.2	194.5	n.a.	33.2	19 020	1.0
Latvia	6 205	2.5	40.3	-1.1	26.6	2 270	-13.7
Liechtenstein	16	n.s.	n.s.	n.a.	78.1	n.a.	n.a.
Lithuania	6 480	3.7	57.1	-0.2	27.0	1 900	-9.7
Malta	32	0.4	1 250.0	0.6	10.2	n.a.	n.a.
Netherlands	3 392	15.7	462.9	0.5	10.9	24 000	1.8
Norway	30 683	4.4	14.3	0.4	26.4	31 250	3.5
Poland	30 442	38.6	126.8	0.1	35.6	2 790	2.4
Portugal	9 150	9.8	107.1	-0.1	63.5	9 740	0.8
Republic of Moldova	3 297	4.4	133.5	0.1	47.0	920	n.a.
Romania	23 034	22.6	98.1	-0.2	43.2	1 480	-1.4
Russian Federation	1 688 850	147.7	8.7	-0.3	23.4	2 240	-9.8
Slovakia	4 808	5.4	112.3	0.1	40.3	2 950	-2.8
Slovenia	2 012	1.9	94.4	-0.1	48.2	8 200	n.a.
Spain	49 944	39.7	79.5	0.1	23.1	13 580	1.1
Sweden	41 162	8.8	21.4	0.2	16.8	23 750	-0.1
Switzerland	3 955	7.3	184.6	0.7	38.4	40 630	0.1
The FYR of Macedonia	2 543	2.2	86.5	0.7	39.3	860	n.a.
Ukraine	57 935	51.4	88.7	-0.4	28.9	1 630	-14.3

Note: The regional breakdown reflects geographical rather than economic or political groupings.

Country	Land area	Population				Economic indicators	
	Total 1996 *(thousand ha)*	Total 1997 *(million)*	Density 1997 *(population/ km²)*	Annual rate of change 1995-2000 *(%)*	Rural 1997 *(%)*	GNP per caput 1995 *(US$)*	Annual growth rate of GDP 1990-1995 *(%)*
United Kingdom	24 160	58.4	241.7	0.1	10.8	18 700	1.4
Yugoslavia	10 200	10.3	101.0	0.5	42.1	n.a.	n.a.
NORTH AND CENTRAL AMERICA							
Antigua and Barbuda	44	0.1	227.3	n.a.	62.7	n.a.	n.a.
Bahamas	1 001	0.3	30.0	1.6	12.8	11 940	n.a.
Barbados	43	0.3	697.7	0.3	51.9	6 560	n.a.
Belize	2 280	0.2	8.8	2.5	53.6	2 630	n.a.
Bermuda	5	0.1	2 000.0	n.a.	0	n.a.	n.a.
British Virgin Islands	15	n.s.	n.s.	n.a.	45.0	n.a.	n.a.
Canada	922 097	29.9	3.2	0.9	23.2	19 380	1.8
Cayman Islands	26	n.s.	n.s.	n.a.	0	n.a.	n.a.
Costa Rica	5 106	3.6	70.5	2.1	49.7	2 610	5.1
Cuba	10 982	11.1	101.1	0.4	23.3	n.a.	n.a.
Dominica	75	0.1	133.3	n.a.	31.0	2 990	n.a.
Dominican Republic	4 838	8.1	167.4	1.6	36.8	1 460	3.9
El Salvador	2 072	5.9	284.7	2.2	54.3	1 610	6.3
Greenland	34 170	n.s.	n.s.	n.a.	18.6	n.a.	n.a.
Grenada	34	0.1	294.1	n.a.	63.4	2 980	n.a.
Guadeloupe	169	0.4	236.7	1.5	0.5	n.a.	n.a.
Guatemala	10 843	11.2	103.3	2.8	60.5	1 340	4.0
Haiti	2 756	7.4	268.5	1.9	67.0	250	-6.5
Honduras	11 189	6.0	53.6	2.8	55.0	600	3.5
Jamaica	1 083	2.5	230.8	0.9	45.3	1 510	2.9
Martinique	106	0.4	377.4	1.0	5.9	n.a.	n.a.
Mexico	190 869	94.3	49.4	1.6	26.2	3 320	1.1
Montserrat	10	n.s.	n.s.	n.a.	81.8	n.a.	n.a.
Netherlands Antilles	80	0.2	250.0	0.7	30.5	n.a.	n.a.
Nicaragua	12 140	4.3	35.4	2.6	36.8	380	1.1
Panama	7 443	2.7	36.3	1.6	43.5	2 750	6.3
Puerto Rico	887	3.8	428.4	0.9	25.9	n.a.	3.0
Saint Kitts and Nevis	36	n.s.	n.s.	n.a.	65.9	5 170	n.a.
Saint Lucia	61	0.1	163.9	n.a.	62.3	3 370	n.a.
Saint Pierre and Miquelon	23	n.s.	n.s.	n.a.	0	n.a.	n.a.
Saint Vincent and Grenadines	39	0.1	256.4	n.a.	49.1	2 280	n.a.
Trinidad and Tobago	513	1.3	253.4	0.8	27.4	3 770	1.0
United States of America	915 912	271.6	29.7	0.8	23.4	26 980	2.6
United States Virgin Islands	34	0.1	294.1	n.a.	54.7	n.a.	n.a.
SOUTH AMERICA							
Argentina	273 669	35.7	13.0	1.3	11.4	8 030	5.7
Bolivia	108 438	7.8	7.2	2.3	37.8	800	3.8
Brazil	845 651	163.1	19.3	1.2	20.5	3 640	2.7
Chile	74 880	14.6	19.5	1.4	15.8	4 160	7.3
Colombia	103 870	37.1	35.7	1.7	26.5	1 910	4.6
Ecuador	27 684	11.9	43.0	2.0	39.7	1 390	3.4
French Guiana	8 815	0.2	2.3	n.a.	23.1	n.a.	n.a.
Guyana	19 685	0.8	4.1	1.0	63.5	590	n.a.
Paraguay	39 730	5.1	12.8	2.6	46.2	1 690	3.1
Peru	128 000	24.4	19.1	1.7	28.4	2 310	5.3
Suriname	15 600	0.4	2.6	1.2	49.7	880	n.a.
Uruguay	17 481	3.2	18.3	0.6	9.3	5 170	4.0
Venezuela	88 205	22.8	25.8	2.0	13.6	3 020	2.4

Note: The regional breakdown reflects geographical rather than economic or political groupings.

TABLE 2
Forest cover, 1995

Country	Land area *(thousand ha)*	Total forest 1995			Natural forest area 1995 *(thousand ha)*
		Area *(thousand ha)*	Percentage of land area	Area per caput *(ha)*	
AFRICA	**2 936 960**	**520 237**	**17.7**	**0.7**	**515 455**
Burkina Faso	27 360	4 271	15.6	0.4	4 251
Cape Verde	403	47	11.7	0.1	6
Chad	125 920	11 025	8.8	1.7	11 021
Gambia	1 000	91	9.1	0	90
Guinea-Bissau	2 812	2 309	82.1	2.2	2 308
Mali	122 019	11 585	9.5	1.1	11 571
Mauritania	102 522	556	0.5	0.2	554
Niger	126 670	2 562	2.0	0.3	2 550
Senegal	19 253	7 381	38.3	0.9	7 269
West Sahelian Africa	**527 959**	**39 827**	**7.5**	**0.8**	**39 620**
Djibouti	2 318	22	0.9	n.s.	22
Eritrea	10 100	282	2.8	0	233
Ethiopia	100 000	13 579	13.6	0.2	13 439
Kenya	56 914	1 292	2.3	n.s.	1 174
Somalia	62 734	754	1.2	0	750
Sudan	237 600	41 613	17.5	1.5	41 410
East Sahelian Africa	**469 666**	**57 542**	**12.3**	**0.5**	**57 028**
Benin	11 062	4 625	41.8	0.9	4 611
Côte d'Ivoire	31 800	5 469	17.2	0.4	5 403
Ghana	22 754	9 022	39.7	0.5	8 969
Guinea	24 572	6 367	25.9	1.0	6 363
Liberia	9 632	4 507	46.8	1.5	4 501
Nigeria	91 077	13 780	15.1	0.1	13 629
Sierra Leone	7 162	1 309	18.3	0.3	1 303
Togo	5 439	1 245	22.9	0.3	1 224
West moist Africa	**203 498**	**46 324**	**22.8**	**0.3**	**46 003**
Burundi	2 568	317	12.3	n.s.	225
Cameroon	46 540	19 598	42.1	1.5	19 582
Central African Republic	62 298	29 930	48.0	9.0	29 924
Congo	34 150	19 537	57.2	7.5	19 500
Congo, Democratic Republic of	226 705	109 245	48.2	2.5	109 203
Equatorial Guinea	2 805	1 781	63.5	4.5	1 778
Gabon	25 767	17 859	69.3	13.5	17 838
Rwanda	2 467	250	10.1	n.s.	162
Sao Tome and Principe	96	56	58.3	0.4	56
Uganda	19 965	6 104	30.6	0.3	6 084
Central Africa	**423 361**	**204 677**	**48.3**	**2.0**	**204 352**
Angola	124 670	22 200	17.8	2.0	22 080
Botswana	56 673	13 917	24.6	9.4	13 916
Malawi	9 408	3 339	35.5	0.3	3 213
Mozambique	78 409	16 862	21.5	1.1	16 834
Namibia	82 329	12 374	15.0	8.0	12 374
Saint Helena	31	1	3.2	0.2	0
United Republic of Tanzania	88 359	32 510	36.8	1.1	32 356

Note: The regional breakdown reflects geographical rather than economic or political groupings.

Country	Land area (thousand ha)	Total forest 1995			Natural forest area 1995 (thousand ha)
		Area (thousand ha)	Percentage of land area	Area per caput (ha)	
Zambia	74 339	31 398	42.2	3.3	31 355
Zimbabwe	38 685	8 710	22.5	0.8	8 626
Tropical southern Africa	**552 903**	**141 311**	**25.6**	**1.5**	**140 754**
Comoros	223	9	4.0	n.s.	9
Madagascar	58 154	15 106	26.0	1.0	14 889
Mauritius	203	12	5.9	n.s.	3
Réunion	250	89	35.6	0.1	82
Seychelles	45	4	8.9	0	3
Insular East Africa	**58 875**	**15 220**	**25.9**	**0.9**	**14 986**
Total tropical Africa	**2 236 262**	**504 901**	**22.6**	**0.9**	**502 743**
Algeria	238 174	1 861	0.8	0	1 376
Egypt	99 545	34	n.s.	n.s.	0
Libyan Arab Jamahiriya	175 954	400	0.2	0	190
Morocco	44 630	3 835	8.6	0.1	3 514
Tunisia	15 536	555	3.6	0	354
North Africa	**573 839**	**6 685**	**1.2**	**0**	**5 434**
Lesotho	3 035	6	0.2	n.s.	0
South Africa	122 104	8 499	7.0	0.2	7 204
Swaziland	1 720	146	8.5	0.2	74
Non-tropical southern Africa	**126 859**	**8 651**	**6.8**	**0.2**	**7 278**
Total non-tropical Africa	**700 698**	**15 336**	**2.2**	**0**	**12 712**
ASIA	**3 073 436**	**503 001**	**16.4**	**0.1**	**n.a.**
Bangladesh	13 017	1 010	7.8	n.s.	700
Bhutan	4 700	2 756	58.6	1.7	2 748
India	297 319	65 005	21.9	0	50 385
Maldives	30	n.a.	n.a.	n.a.	n.a.
Nepal	14 300	4 822	33.7	0.2	4 766
Pakistan	77 088	1 748	2.3	n.s.	1 580
Sri Lanka	6 463	1 796	27.8	0	1 657
South Asia	**412 917**	**77 137**	**18.7**	**0**	**61 836**
Cambodia	17 652	9 830	55.7	1.0	9 823
Laos	23 080	12 435	53.9	2.5	12 431
Myanmar	65 755	27 151	41.3	0.6	26 875
Thailand	51 089	11 630	22.8	0.2	11 101
Viet Nam	32 549	9 117	28.0	0.1	7 647
Continental Southeast Asia	**190 125**	**70 163**	**36.9**	**0.4**	**67 877**
Brunei Darussalam	527	434	82.4	1.5	434
Indonesia	181 157	109 791	60.6	0.6	103 666
Malaysia	32 855	15 471	47.1	0.8	15 371
Philippines	29 817	6 766	22.7	0.1	6 563
Singapore	61	4	6.6	n.s.	4
Insular Southeast Asia	**244 417**	**132 466**	**54.2**	**0.5**	**126 038**
Total tropical Asia	**847 459**	**279 766**	**33.0**	**0.2**	**255 751**

Note: The regional breakdown reflects geographical rather than economic or political groupings.

Country	Land area (thousand ha)	Total forest 1995			Natural forest area 1995 (thousand ha)
		Area (thousand ha)	Percentage of land area	Area per caput (ha)	
Afghanistan	65 209	1 398	2.1	0	1 390
Armenia	2 820	334	11.8	0	n.ap.
Azerbaijan	8 660	990	11.4	0.1	n.ap.
Bahrain	69	0	0	0	0
Cyprus	924	140	15.2	0.2	n.ap.
Georgia	6 970	2 988	42.9	0.5	n.ap.
Iran, Islamic Republic of	162 200	1 544	1.0	n.s.	1 465
Iraq	43 737	83	0.2	n.s.	69
Israel	2 062	102	4.9	n.s.	n.ap.
Jordan	8 893	45	0.5	n.s.	22
Kazakhstan	267 073	10 504	3.9	0.6	n.ap.
Kuwait	1 782	5	0.3	n.s.	0
Kyrgyzstan	19 180	730	3.8	0.2	n.ap.
Lebanon	1 023	52	5.1	n.s.	39
Oman	21 246	0	0	0	0
Qatar	1 100	0	0	0	0
Saudi Arabia	214 969	222	0.1	n.s.	221
Syrian Arab Republic	18 378	219	1.2	n.s.	92
Tajikistan	14 060	410	2.9	0	n.ap.
Turkey	76 963	8 856	11.5	0.1	n.ap.
Turkmenistan	46 993	3 754	8.0	0.9	n.ap.
United Arab Emirates	8 360	60	0.7	n.s.	0
Uzbekistan	41 424	9 119	22.0	0.4	n.ap.
Yemen	52 797	9	n.s.	n.s.	9
Western and Central Asia	**1 086 892**	**41 564**	**3.8**	**0.1**	**n.ap.**
China	932 641	133 323	14.3	0.1	99 523
China – Hong Kong SAR	99	n.a.	n.a.	n.a.	n.a.
DPR of Korea	12 041	6 170	51.2	0.3	4 700
Japan	37 652	25 146	66.8	0.2	n.ap.
Macau	2	n.a.	n.a.	n.a.	n.a.
Mongolia	156 650	9 406	6.0	3.9	9 406
Republic of Korea	9 873	7 626	77.2	0.2	6 226
East Asia	**1 139 085**	**181 671**	**15.9**	**0.1**	**n.ap.**
Total temperate Asia	**2 225 977**	**223 235**	**10.0**	**0.1**	**n.ap.**
OCEANIA	**849 084**	**90 695**	**10.7**	**3.2**	**n.ap.**
American Samoa	20	0	0	0	0
Cook Islands	23	n.a.	n.a.	n.a.	n.a.
Fiji	1 827	835	45.7	1.1	757
French Polynesia	366	0	0	0	0
Guam	55	0	0	0	0
Kiribati	73	0	0	0	0
New Caledonia	1 828	698	38.2	3.9	689
Niue	26	6	23.1	3.0	6
Pacific Islands (Trust Terr.)	178	0	0	0	0
Papua New Guinea	45 286	36 939	81.6	8.6	36 909
Samoa	283	136	48.1	0.8	127
Solomon Islands	2 799	2 389	85.4	6.3	2 371
Tonga	72	0	0	0	0
Vanuatu	1 219	900	73.8	5.3	893
Total tropical Oceania	**54 055**	**41 903**	**77.5**	**6.3**	**41 752**

Note: The regional breakdown reflects geographical rather than economic or political groupings.

Country	Land area (thousand ha)	Total forest 1995 Area (thousand ha)	Total forest 1995 Percentage of land area	Total forest 1995 Area per caput (ha)	Natural forest area 1995 (thousand ha)
Australia	768 230	40 908	5.3	2.3	n.ap.
New Zealand	26 799	7 884	29.4	2.2	n.ap.
Total temperate Oceania	**795 029**	**48 792**	**6.1**	**2.3**	**n.ap.**
EUROPE	**2 260 128**	**933 326**	**41.3**	**1.3**	**n.ap.**
Finland	30 459	20 029	65.8	3.9	n.ap.
Iceland	10 025	11	0.1	n.s.	n.ap.
Norway	30 683	8 073	26.3	1.9	n.ap.
Sweden	41 162	24 425	59.3	2.8	n.ap.
Northern Europe	**112 329**	**52 538**	**46.8**	**2.8**	**n.ap.**
Austria	8 273	3 877	46.9	0.5	n.ap.
Belgium and Luxembourg	3 282	709	21.6	0	n.ap.
Denmark	4 243	417	9.8	0	n.ap.
France	55 010	15 034	27.3	0.3	n.ap.
Germany	34 927	10 740	30.7	0.1	n.ap.
Greece	12 890	6 513	50.5	0.6	n.ap.
Ireland	6 889	570	8.3	0.2	n.ap.
Italy	29 406	6 496	22.1	0.1	n.ap.
Liechtenstein	16	6	37.5	0.2	n.ap.
Malta	32	0	0	0	n.ap.
Netherlands	3 392	334	9.8	n.s.	n.ap.
Portugal	9 150	2 875	31.4	0.3	n.ap.
Spain	49 944	8 388	16.8	0.2	n.ap.
Switzerland	3 955	1 130	28.6	0.2	n.ap.
United Kingdom	24 160	2 390	9.9	n.s.	n.ap.
Western Europe	**245 569**	**59 479**	**24.2**	**0.2**	**n.ap.**
Albania	2 740	1 046	38.2	0.3	n.ap.
Belarus	20 748	7 372	35.5	0.7	n.ap.
Bosnia and Herzegovina	5 100	2 710	53.1	0.8	n.ap.
Bulgaria	11 055	3 240	29.3	0.4	n.ap.
Croatia	5 592	1 825	32.6	0.4	n.ap.
Czech Republic	7 728	2 630	34.0	0.3	n.ap.
Estonia	4 227	2 011	47.6	1.3	n.ap.
Hungary	9 234	1 719	18.6	0.2	n.ap.
Latvia	6 205	2 882	46.4	1.1	n.ap.
Lithuania	6 480	1 976	30.5	0.5	n.ap.
Poland	30 442	8 732	28.7	0.2	n.ap.
Republic of Moldova	3 297	357	10.8	0	n.ap.
Romania	23 034	6 246	27.1	0.3	n.ap.
Russian Federation	1 688 850	763 500	45.2	5.2	n.ap.
Slovakia	4 808	1 989	41.4	0.4	n.ap.
Slovenia	2 012	1 077	53.5	0.6	n.ap.
The FYR of Macedonia	2 543	988	38.9	0.5	n.ap.
Ukraine	57 935	9 240	15.9	0.2	n.ap.
Yugoslavia	10 200	1 769	17.3	0.2	n.ap.
Eastern Europe	**1 902 230**	**821 309**	**43.2**	**2.4**	**n.ap.**
NORTH AND CENTRAL AMERICA	**2 102 811**	**536 529**	**25.5**	**1.2**	**n.ap.**
Bermuda	5	0	0	0	0
Canada	922 097	244 571	26.5	8.3	n.ap.

Note: The regional breakdown reflects geographical rather than economic or political groupings.

Country	Land area (thousand ha)	Total forest 1995			Natural forest area 1995 (thousand ha)
		Area (thousand ha)	Percentage of land area	Area per caput (ha)	
Saint Pierre and Miquelon	23	n.a.	n.a.	n.a.	n.a.
United States of America	915 912	212 515	23.2	0.8	n.ap.
Total temperate North and Central America	**1 838 037**	**457 086**	**24.9**	**1.6**	**n.ap.**
Belize	2 280	1 962	86.1	9.1	1 960
Costa Rica	5 106	1 248	24.4	0.4	1 220
El Salvador	2 072	105	5.1	n.s.	101
Guatemala	10 843	3 841	35.4	0.4	3 813
Honduras	11 189	4 115	36.8	0.7	4 112
Mexico	190 869	55 387	29.0	0.6	55 278
Nicaragua	12 140	5 560	45.8	1.3	5 546
Panama	7 443	2 800	37.6	1.1	2 794
Central America and Mexico	**241 942**	**75 018**	**31.0**	**0.6**	**74 824**
Antigua and Barbuda	44	9	20.5	0.1	9
Bahamas	1 001	158	15.8	0.6	158
Barbados	43	0	0	0	0
British Virgin Islands	15	4	26.7	0.2	4
Cayman Islands	26	0	0	0	0
Cuba	10 982	1 842	16.8	0.2	1 597
Dominica	75	46	61.3	0.6	46
Dominican Republic	4 838	1 582	32.7	0.2	1 575
Grenada	34	4	11.8	n.s.	4
Guadeloupe	169	80	47.3	0.2	80
Haiti	2 756	21	0.8	n.s.	13
Jamaica	1 083	175	16.2	0	160
Martinique	106	38	35.8	0.1	38
Montserrat	10	3	30.0	0.3	3
Netherlands Antilles	80	0	0	0	0
Puerto Rico	887	275	31.0	0	272
Saint Kitts and Nevis	36	11	30.6	0.3	11
Saint Lucia	61	5	8.2	n.s.	5
Saint Vincent and Grenadines	39	11	28.2	0	11
Trinidad and Tobago	513	161	31.4	0.1	148
United States Virgin Islands	34	0	0	0	0
Caribbean	**22 832**	**4 425**	**19.4**	**0.1**	**4 134**
Total tropical and subtropical North and Central America	**264 774**	**79 443**	**30.0**	**0.5**	**78 958**
SOUTH AMERICA	**1 751 708**	**870 594**	**49.7**	**2.7**	**863 315**
Bolivia	108 438	48 310	44.6	6.5	48 282
Brazil	845 651	551 139	65.2	3.4	546 239
Colombia	103 870	52 988	51.0	1.5	52 862
Ecuador	27 684	11 137	40.2	1.0	11 092
French Guiana	8 815	7 990	90.6	54.4	7 990
Guyana	19 685	18 577	94.4	22.2	18 569
Paraguay	39 730	11 527	29.0	2.3	11 518
Peru	128 000	67 562	52.8	2.8	67 378
Suriname	15 600	14 721	94.4	34.8	14 713
Venezuela	88 205	43 995	49.9	2.0	43 742
Total tropical South America	**1 385 678**	**827 946**	**59.8**	**3.1**	**822 385**

Note: The regional breakdown reflects geographical rather than economic or political groupings.

Country	Land area *(thousand ha)*	Total forest 1995			Natural forest area 1995 *(thousand ha)*
		Area *(thousand ha)*	Percentage of land area	Area per caput *(ha)*	
Argentina	273 669	33 942	12.4	1.0	33 395
Chile	74 880	7 892	10.5	0.6	6 877
Uruguay	17 481	814	4.7	0.3	658
Total temperate South America	**366 030**	**42 648**	**11.7**	**0.8**	**40 930**
WORLD TOTAL	**13 048 410**	**3 454 382**	**26.6**	**0.6**	**n.ap.**

Note: The regional breakdown reflects geographical rather than economic or political groupings.

TABLE 3
Change in forest cover, 1990-1995

Country	Total forest 1990 (thousand ha)	Total forest 1995 (thousand ha)	Total change 1990-1995 (thousand ha)	Annual change (thousand ha)	Annual change rate (%)
AFRICA	**538 978**	**520 237**	**-18 741**	**-3 748**	**-0.7**
Burkina Faso	4 431	4 271	-160	-32	-0.7
Cape Verde	16	47	31	6	24.0
Chad	11 496	11 025	-471	-94	-0.8
Gambia	95	91	-4	0	-0.9
Guinea-Bissau	2 361	2 309	-52	-10	-0.4
Mali	12 154	11 585	-569	-114	-1.0
Mauritania	556	556	0	0	0
Niger	2 562	2 562	0	0	0
Senegal	7 629	7 381	-248	-50	-0.7
West Sahelian Africa	**41 300**	**39 827**	**-1 473**	**-295**	**-0.7**
Djibouti	22	22	0	0	0
Eritrea	282	282	0	0	0
Ethiopia	13 891	13 579	-312	-62	-0.5
Kenya	1 309	1 292	-17	-3	-0.3
Somalia	760	754	-6	-1	-0.2
Sudan	43 376	41 613	-1 763	-353	-0.8
East Sahelian Africa	**59 640**	**57 542**	**-2 098**	**-420**	**-0.7**
Benin	4 923	4 625	-298	-60	-1.2
Côte d'Ivoire	5 623	5 469	-154	-31	-0.6
Ghana	9 608	9 022	-586	-117	-1.3
Guinea	6 741	6 367	-374	-75	-1.1
Liberia	4 641	4 507	-134	-27	-0.6
Nigeria	14 387	13 780	-607	-121	-0.9
Sierra Leone	1 522	1 309	-213	-43	-3.0
Togo	1 338	1 245	-93	-19	-1.4
West moist Africa	**48 783**	**46 324**	**-2 459**	**-492**	**-1.0**
Burundi	324	317	-7	-1	-0.4
Cameroon	20 244	19 598	-646	-129	-0.6
Central African Republic	30 571	29 930	-641	-128	-0.4
Congo	19 745	19 537	-208	-42	-0.2
Congo, Democratic Republic of	112 946	109 245	-3 701	-740	-0.7
Equatorial Guinea	1 829	1 781	-48	-10	-0.5
Gabon	18 314	17 859	-455	-91	-0.5
Rwanda	252	250	-2	n.s.	-0.2
Sao Tome and Principe	56	56	0	0	0
Uganda	6 400	6 104	-296	-59	-0.9
Central Africa	**210 681**	**204 677**	**-6 004**	**-1 201**	**-0.6**
Angola	23 385	22 200	-1 185	-237	-1.0
Botswana	14 271	13 917	-354	-71	-0.5
Malawi	3 612	3 339	-273	-55	-1.6
Mozambique	17 443	16 862	-581	-116	-0.7
Namibia	12 584	12 374	-210	-42	-0.3
Saint Helena	1	1	0	0	0
United Republic of Tanzania	34 123	32 510	-1 613	-323	-1.0
Zambia	32 720	31 398	-1 322	-264	-0.8
Zimbabwe	8 960	8 710	-250	-50	-0.6
Tropical southern Africa	**147 099**	**141 311**	**-5 788**	**-1 158**	**-0.8**

Note: The regional breakdown reflects geographical rather than economic or political groupings.

Country	Total forest 1990 *(thousand ha)*	Total forest 1995 *(thousand ha)*	Total change 1990-1995 *(thousand ha)*	Annual change *(thousand ha)*	Annual change rate *(%)*
Comoros	12	9	-3	0	-5.6
Madagascar	15 756	15 106	-650	-130	-0.8
Mauritius	12	12	0	0	0
Réunion	89	89	0	0	0
Seychelles	4	4	0	0	0
Insular East Africa	**15 873**	**15 220**	**-653**	**-131**	**-0.8**
Total tropical Africa	**523 376**	**504 901**	**-18 475**	**-3 695**	**-0.7**
Algeria	1 978	1 861	-117	-23	-1.2
Egypt	34	34	0	0	0
Libyan Arab Jamahiriya	400	400	0	0	0
Morocco	3 894	3 835	-59	-12	-0.3
Tunisia	570	555	-15	-3	-0.5
North Africa	**6 876**	**6 685**	**-191**	**-38**	**-0.6**
Lesotho	6	6	0	0	0
South Africa	8 574	8 499	-75	-15	-0.2
Swaziland	146	146	0	0	0
Non-tropical southern Africa	**8 726**	**8 651**	**-75**	**-15**	**-0.2**
Total non-tropical Africa	**15 602**	**15 336**	**-266**	**-53**	**-0.3**
ASIA	**517 505**	**503 001**	**-14 504**	**-2 901**	**-0.6**
Bangladesh	1 054	1 010	-44	-9	-0.8
Bhutan	2 803	2 756	-47	-9	-0.3
India	64 969	65 005	36	7	n.s.
Maldives	n.a.	n.a.	n.a.	n.a.	n.a.
Nepal	5 096	4 822	-274	-55	-1.1
Pakistan	2 023	1 748	-275	-55	-2.9
Sri Lanka	1 897	1 796	-101	-20	-1.1
South Asia	**77 842**	**77 137**	**-705**	**-141**	**-0.2**
Cambodia	10 649	9 830	-819	-164	-1.6
Laos	13 177	12 435	-742	-148	-1.2
Myanmar	29 088	27 151	-1 937	-387	-1.4
Thailand	13 277	11 630	-1 647	-329	-2.6
Viet Nam	9 793	9 117	-676	-135	-1.4
Continental Southeast Asia	**75 984**	**70 163**	**-5 821**	**-1 164**	**-1.6**
Brunei Darussalam	448	434	-14	-3	-0.6
Indonesia	115 213	109 791	-5 422	-1 084	-1.0
Malaysia	17 472	15 471	-2 001	-400	-2.4
Philippines	8 078	6 766	-1 312	-262	-3.5
Singapore	4	4	0	0	0
Insular Southeast Asia	**141 215**	**132 466**	**-8 749**	**-1 750**	**-1.3**
Total tropical Asia	**295 041**	**279 766**	**-15 275**	**-3 055**	**-1.1**
Afghanistan	1 990	1 398	-592	-118	-6.8
Armenia	292	334	42	8	2.7
Azerbaijan	990	990	0	0	0
Bahrain	0	0	0	0	0
Cyprus	140	140	0	0	0
Georgia	2 988	2 988	0	0	0
Iran, Islamic Republic of	1 686	1 544	-142	-28	-1.7
Iraq	83	83	0	0	0

Note: The regional breakdown reflects geographical rather than economic or political groupings.

Country	Total forest 1990 *(thousand ha)*	Total forest 1995 *(thousand ha)*	Total change 1990-1995 *(thousand ha)*	Annual change *(thousand ha)*	Annual change rate *(%)*
Israel	102	102	0	0	0
Jordan	51	45	-6	-1	-2.5
Kazakhstan	9 540	10 504	964	193	1.9
Kuwait	5	5	0	0	0
Kyrgyzstan	730	730	0	0	0
Lebanon	78	52	-26	-5	-7.8
Oman	0	0	0	0	0
Qatar	0	0	0	0	0
Saudi Arabia	231	222	-9	-2	-0.8
Syrian Arab Republic	245	219	-26	-5	-2.2
Tajikistan	410	410	0	0	0
Turkey	8 856	8 856	0	0	0
Turkmenistan	3 754	3 754	0	0	0
United Arab Emirates	60	60	0	0	0
Uzbekistan	7 989	9 119	1 130	226	2.7
Yemen	9	9	0	0	0
Western and Central Asia	**40 229**	**41 564**	**1 335**	**267**	**0.7**
China	133 756	133 323	-433	-87	-0.1
China – Hong Kong SAR	n.a.	n.a.	n.a.	n.a.	n.a.
DPR of Korea	6 170	6 170	0	0	0
Japan	25 212	25 146	-66	-13	-0.1
Macau	n.a.	n.a.	n.a.	n.a.	n.a.
Mongolia	9 406	9 406	0	0	0
Republic of Korea	7 691	7 626	-65	-13	-0.2
East Asia	**182 235**	**181 671**	**-564**	**-113**	**-0.1**
Total temperate Asia	**222 464**	**223 235**	**771**	**154**	**0**
OCEANIA	**91 149**	**90 695**	**-454**	**-91**	**-0.1**
American Samoa	0	0	0	0	0
Cook Islands	n.a.	n.a.	n.a.	n.a.	n.a.
Fiji	853	835	-18	-4	-0.4
French Polynesia	0	0	0	0	0
Guam	0	0	0	0	0
Kiribati	0	0	0	0	0
New Caledonia	701	698	-3	-1	-0.1
Niue	6	6	0	0	0
Pacific Islands (Trust Terr.)	0	0	0	0	0
Papua New Guinea	37 605	36 939	-666	-133	-0.4
Samoa	144	136	-8	-2	-1.1
Solomon Islands	2 412	2 389	-23	-5	-0.2
Tonga	0	0	0	0	n.a.
Vanuatu	938	900	-38	-8	-0.8
Total tropical Oceania	**42 659**	**41 903**	**-756**	**-151**	**-0.4**
Australia	40 823	40 908	85	17	n.s.
New Zealand	7 667	7 884	217	43	0.6
Total temperate Oceania	**48 490**	**48 792**	**302**	**60**	**0.1**
EUROPE	**930 732**	**933 326**	**2 594**	**519**	**0**
Finland	20 112	20 029	-83	-17	-0.1
Iceland	11	11	0	0	0
Norway	7 938	8 073	135	27	0.3
Sweden	24 437	24 425	-12	-2	n.s.
Northern Europe	**52 498**	**52 538**	**40**	**8**	**n.s.**

Note: The regional breakdown reflects geographical rather than economic or political groupings.

Country	Total forest 1990 *(thousand ha)*	Total forest 1995 *(thousand ha)*	Total change 1990-1995 *(thousand ha)*	Annual change *(thousand ha)*	Annual change rate *(%)*
Austria	3 877	3 877	0	0	0
Belgium and Luxembourg	709	709	0	0	0
Denmark	417	417	0	0	0
France	14 230	15 034	804	161	1.1
Germany	10 740	10 740	0	0	0
Greece	5 809	6 513	704	141	2.3
Ireland	500	570	70	14	2.7
Italy	6 467	6 496	29	6	0
Liechtenstein	6	6	0	0	0
Malta	0	0	0	0	0
Netherlands	334	334	0	0	0
Portugal	2 755	2 875	120	24	0.9
Spain	8 388	8 388	0	0	0
Switzerland	1 130	1 130	0	0	0
United Kingdom	2 326	2 390	64	13	0.5
Western Europe	**57 688**	**59 479**	**1 791**	**358**	**0.6**
Albania	1 046	1 046	0	0	0
Belarus	7 028	7 372	344	69	1.0
Bosnia and Herzegovina	2 710	2 710	0	0	0
Bulgaria	3 237	3 240	3	0	n.s.
Croatia	2 629	2 630	1	n.s.	n.s.
Czech Republic	1 913	2 011	98	20	1.0
Estonia	1 675	1 719	44	9	0.5
Hungary	2 757	2 882	125	25	0.9
Latvia	1 825	1 825	0	0	0
Lithuania	1 920	1 976	56	11	0.6
Poland	8 672	8 732	60	12	0.1
Republic of Moldova	357	357	0	0	0
Romania	6 252	6 246	-6	-1	n.s.
Russian Federation	763 500	763 500	n.a.	n.a.	n.a.
Slovakia	1 977	1 989	12	2	0.1
Slovenia	1 077	1 077	0	0	0
The FYR of Macedonia	989	988	-1	n.s.	n.s.
Ukraine	9 213	9 240	27	5	0
Yugoslavia	1 769	1 769	0	0	0
Eastern Europe	**820 546**	**821 309**	**763**	**153**	**0**
NORTH AND CENTRAL AMERICA	**537 898**	**536 529**	**-1 369**	**-274**	**-0.1**
Bermuda	0	0	0	0	0
Canada	243 698	244 571	873	175	0
Saint Pierre and Miquelon	n.a.	n.a.	n.a.	n.a.	n.a.
United States of America	209 572	212 515	2 943	589	0.3
Total temperate North and Central America	**453 270**	**457 086**	**3 816**	**763**	**0.2**
Belize	1 995	1 962	-33	-7	-0.3
Costa Rica	1 455	1 248	-207	-41	-3.0
El Salvador	124	105	-19	-4	-3.3
Guatemala	4 253	3 841	-412	-82	-2.0
Honduras	4 626	4 115	-511	-102	-2.3
Mexico	57 927	55 387	-2 540	-508	-0.9
Nicaragua	6 314	5 560	-754	-151	-2.5
Panama	3 118	2 800	-318	-64	-2.1
Central America and Mexico	**79 812**	**75 018**	**-4 794**	**-959**	**-1.2**

Note: The regional breakdown reflects geographical rather than economic or political groupings.

Country	Total forest 1990 *(thousand ha)*	Total forest 1995 *(thousand ha)*	Total change 1990-1995 *(thousand ha)*	Annual change *(thousand ha)*	Annual change rate *(%)*
Antigua and Barbuda	9	9	0	0	0
Bahamas	180	158	-22	-4	-2.6
Barbados	0	0	0	0	0
British Virgin Islands	5	4	-1	0	-4.4
Cayman Islands	0	0	0	0	0
Cuba	1 960	1 842	-118	-24	-1.2
Dominica	46	46	0	0	0
Dominican Republic	1 714	1 582	-132	-26	-1.6
Grenada	4	4	0	0	0
Guadeloupe	87	80	-7	-1	-1.7
Haiti	25	21	-4	0	-3.4
Jamaica	254	175	-79	-16	-7.2
Martinique	40	38	-2	n.s.	-1.0
Montserrat	3	3	0	0	0
Netherlands Antilles	0	0	0	0	0
Puerto Rico	287	275	-12	-2	-0.9
Saint Kitts and Nevis	11	11	0	0	0
Saint Lucia	6	5	-1	0	-3.6
Saint Vincent and Grenadines	11	11	0	0	0
Trinidad and Tobago	174	161	-13	-3	-1.5
United States Virgin Islands	0	0	0	0	0
Caribbean	**4 816**	**4 425**	**-391**	**-78**	**-1.7**
Total tropical and subtropical North and Central America	**84 628**	**79 443**	**-5 185**	**-1 037**	**-1.3**
SOUTH AMERICA	**894 466**	**870 594**	**-23 872**	**-4 774**	**-0.5**
Bolivia	51 217	48 310	-2 907	-581	-1.2
Brazil	563 911	551 139	-12 772	-2 554	-0.5
Colombia	54 299	52 988	-1 311	-262	-0.5
Ecuador	12 082	11 137	-945	-189	-1.6
French Guiana	7 994	7 990	-4	0	n.s.
Guyana	18 620	18 577	-43	-9	n.s.
Paraguay	13 160	11 527	-1 633	-327	-2.6
Peru	68 646	67 562	-1 084	-217	-0.3
Suriname	14 782	14 721	-61	-12	-0.1
Venezuela	46 512	43 995	-2 517	-503	-1.1
Total tropical South America	**851 223**	**827 946**	**-23 277**	**-4 655**	**-0.6**
Argentina	34 389	33 942	-447	-89	-0.3
Chile	8 038	7 892	-146	-29	-0.4
Uruguay	816	814	-2	n.s.	n.s.
Total temperate South America	**43 243**	**42 648**	**-595**	**-119**	**-0.3**
WORLD TOTAL	**3 510 728**	**3 454 382**	**-56 346**	**-11 269**	**-0.3**

Note: The regional breakdown reflects geographical rather than economic or political groupings.

TABLE 4

Production, trade and consumption of forest products, 1996

Country	Fuelwood and charcoal *(thousand m³)*				Industrial roundwood *(thousand m³)*				Sawnwood *(thousand m³)*			
	Production	Imports	Exports	Consumption	Production	Imports	Exports	Consumption	Production	Imports	Exports	Consumption
AFRICA	**519 884**	**399**	**547**	**519 736**	**67 931**	**676**	**7 439**	**61 168**	**9 147**	**4 599**	**1 351**	**12 395**
Algeria	2 176	0	0	2 176	395	47	0	442	13	371	0	384
Angola	6 184	0	0	6 184	1 031	0	0	1 031	5	3	0	8
Benin	5 753	0	0	5 753	329	0	0	329	24	0	2	22
Botswana	1 531	0	0	1 531	101	0	0	101	0	0	0	0
Burkina Faso	9 835	0	0	9 835	466	0	0	466	2	0	0	2
Burundi	5 006	0	0	5 006	43	0	0	43	33	0	0	33
Cameroon	12 692	0	0	12 692	3 364	0	1 307	2 057	1 400	0	316	1 084
Cape Verde	0	0	0	0	0	2	0	2	0	7	0	7
Central African Republic	3 250	0	0	3 250	600	0	42	558	61	0	31	30
Chad	3 993	0	0	3 993	669	0	0	669	2	17	0	20
Comoros	0	0	0	0	0	0	0	0	0	1	0	1
Congo	2 423	0	0	2 423	1 485	0	297	1 188	62	0	37	25
Congo, Democratic Republic of	45 142	0	0	45 142	3 433	0	178	3 255	100	0	44	56
Côte d'Ivoire	11 974	0	6	11 968	3 008	0	332	2 676	706	0	501	206
Djibouti	0	0	0	0	0	0	0	0	0	1	0	1
Egypt	2 625	296	240	2 681	128	113	0	241	0	2 534	2	2 532
Equatorial Guinea	447	0	0	447	364	0	267	97	4	0	4	0
Eritrea	n.a.	n.a.	n.a.	n.a.	n.a.	n.a.	n.a.	n.a.	n.a.	n.a.	n.a.	n.a.
Ethiopia	46 982	0	0	46 982	2 373	0	2	2 371	33	1	0	34
Gabon	2 975	0	0	2 975	1 990	0	2 231	- 240	170	0	12	158
Gambia	1 120	0	0	1 120	113	0	0	113	1	0	0	2
Ghana	25 190	0	0	25 190	1 255	0	150	1 105	604	0	240	364
Guinea	4 273	0	0	4 273	651	0	31	620	85	3	1	87
Guinea-Bissau	422	0	0	422	160	0	16	144	16	0	0	15
Kenya	41 003	0	0	41 003	1 961	0	0	1 961	185	2	2	185
Lesotho	728	0	0	728	0	0	0	0	0	0	0	0
Liberia	5 370	0	0	5 370	993	0	50	943	90	0	0	90
Libyan Arab Jamahiriya	536	0	0	536	117	5	0	122	31	184	0	215
Madagascar	10 746	0	0	10 745	434	2	1	435	69	0	10	59
Malawi	10 153	0	0	10 153	547	0	0	547	45	0	2	43
Mali	6 315	0	0	6 315	420	0	0	420	13	0	0	13
Mauritania	9	0	0	9	6	0	0	6	0	0	0	0
Mauritius	7	0	0	7	8	9	0	17	3	28	0	31
Morocco	1 484	0	0	1 484	941	334	0	1 275	83	540	0	622
Mozambique	17 324	0	0	17 324	1 098	0	16	1 082	42	0	2	40
Namibia	n.a.	n.a.	n.a.	n.a.	n.a.	n.a.	n.a.	n.a.	n.a.	n.a.	n.a.	n.a.
Niger	5 693	0	0	5 693	374	0	0	374	4	0	0	4
Nigeria	105 832	0	101	105 731	8 479	2	11	8 470	2 723	0	25	2 698
Réunion	31	4	0	35	5	2	2	5	2	85	0	87
Rwanda	5 392	0	0	5 392	268	0	4	264	36	0	0	37
Saint Helena	0	0	0	0	0	0	0	0	0	0	0	0
Sao Tome and Principe	0	0	0	0	9	0	0	9	5	0	0	5
Senegal	4 457	0	0	4 457	737	25	0	761	23	30	0	53
Seychelles	0	0	0	0	0	0	0	0	0	0	0	0
Sierra Leone	3 279	0	0	3 279	124	0	0	124	5	0	0	5
Somalia	8 911	0	0	8 911	108	0	0	108	14	0	0	14
South Africa	7 156	97	198	7 055	18 176	77	2 473	15 780	1 574	413	39	1 948
Sudan	14 600	0	0	14 601	2 274	14	0	2 287	45	70	0	115
Swaziland	560	0	0	560	864	0	0	864	75	0	0	75
Togo	2 189	0	0	2 189	245	3	3	245	14	3	0	18
Tunisia	3 417	1	0	3 418	210	38	0	248	20	298	0	319

Note: The regional breakdown reflects geographical rather than economic or political groupings.

Wood-based panels (thousand m³)				Pulp for paper (thousand m³)				Paper and paperboard (thousand m³)				Country
Production	Imports	Exports	Consumption	Production	Imports	Exports	Consumption	Production	Imports	Exports	Consumption	
1 822	**544**	**347**	**2 019**	**2 131**	**331**	**754**	**1 708**	**2 695**	**1 205**	**494**	**3 406**	**AFRICA**
50	57	0	107	21	63	0	84	78	121	0	199	Algeria
11	2	0	13	15	0	0	15	0	1	0	1	Angola
0	0	0	0	0	0	0	0	0	0	0	0	Benin
0	0	0	0	0	0	0	0	0	0	0	0	Botswana
0	0	0	0	0	0	0	0	0	0	0	0	Burkina Faso
0	0	0	0	0	0	0	0	0	3	0	3	Burundi
74	1	39	36	0	0	0	0	5	31	0	36	Cameroon
0	2	0	2	0	0	0	0	0	0	0	0	Cape Verde
2	0	0	2	0	0	0	0	0	0	0	0	Central African Republic
0	2	0	2	0	0	0	0	0	0	0	0	Chad
0	1	0	1	0	0	0	0	0	0	0	0	Comoros
52	0	31	21	0	0	0	0	0	0	0	0	Congo
21	0	5	16	0	0	0	0	3	2	0	5	Congo, Democratic Republic of
272	0	74	199	0	0	0	0	0	52	0	52	Côte d'Ivoire
0	0	0	0	0	0	0	0	0	0	0	0	Djibouti
81	128	7	202	60	64	0	124	221	263	2	483	Egypt
9	0	9	0	0	1	0	1	0	0	0	0	Equatorial Guinea
n.a.	n.a.	n.a.	n.a.	n.a.	n.a.	n.a.	n.a.	n.a.	n.a.	n.a.	n.a.	Eritrea
12	0	0	12	0	5	0	5	7	3	0	10	Ethiopia
27	0	24	4	0	0	0	0	0	1	0	1	Gabon
0	2	0	2	0	0	0	0	0	0	0	0	Gambia
105	0	73	33	0	0	0	0	0	13	0	13	Ghana
0	6	0	6	0	0	0	0	0	2	0	2	Guinea
0	0	0	0	0	0	0	0	0	0	0	0	Guinea-Bissau
52	3	0	55	66	5	0	71	129	20	0	149	Kenya
0	0	0	0	0	0	0	0	0	0	0	0	Lesotho
8	0	0	8	0	0	0	0	0	0	0	0	Liberia
0	30	0	30	0	2	0	2	6	9	0	15	Libyan Arab Jamahiriya
5	0	0	6	3	0	0	3	4	5	0	9	Madagascar
18	0	6	12	0	0	0	0	0	4	0	4	Malawi
0	0	0	0	0	0	0	0	0	2	0	2	Mali
0	0	0	0	0	0	0	0	0	4	0	4	Mauritania
0	26	0	26	0	0	0	0	0	19	0	19	Mauritius
35	57	25	67	104	24	68	60	106	138	4	239	Morocco
0	0	0	0	0	10	0	10	1	0	0	1	Mozambique
n.a.	n.a.	n.a.	n.a.	n.a.	n.a.	n.a.	n.a.	n.a.	n.a.	n.a.	n.a.	Namibia
0	0	0	0	0	0	0	0	0	0	0	0	Niger
115	20	0	135	7	7	0	14	57	55	0	111	Nigeria
0	24	0	23	0	0	0	0	0	15	0	15	Réunion
2	6	0	8	0	0	0	0	0	6	0	6	Rwanda
0	0	0	0	0	0	0	0	0	0	0	0	Saint Helena
0	0	0	0	0	0	0	0	0	0	0	0	Sao Tome and Principe
0	8	0	8	0	0	0	0	0	10	0	10	Senegal
0	1	0	1	0	0	0	0	0	0	0	0	Seychelles
0	0	0	0	0	0	0	0	0	2	0	2	Sierra Leone
0	0	0	0	0	0	0	0	0	0	0	0	Somalia
653	111	23	741	1 547	95	465	1 177	1 871	316	479	1 708	South Africa
2	4	0	5	0	0	0	0	3	3	0	6	Sudan
8	0	0	8	200	0	210	-10	0	0	0	0	Swaziland
0	0	0	0	0	0	0	0	0	1	0	1	Togo
104	37	5	136	12	50	10	52	90	67	3	154	Tunisia

Country	Fuelwood and charcoal *(thousand m^3)*				Industrial roundwood *(thousand m^3)*				Sawnwood *(thousand m^3)*			
	Production	Imports	Exports	Consumption	Production	Imports	Exports	Consumption	Production	Imports	Exports	Consumption
Uganda	15 410	0	0	15 410	2 352	0	0	2 352	83	0	0	83
United Republic of Tanzania	35 556	0	0	35 556	2 203	0	8	2 195	24	0	4	21
Zambia	13 465	0	0	13 465	1 163	0	0	1 163	367	0	2	365
Zimbabwe	6 269	0	0	6 269	1 859	0	15	1 844	250	2	55	196
ASIA	**905 236**	**1 682**	**2 118**	**904 800**	**280 153**	**68 509**	**14 336**	**334 326**	**98 166**	**23 023**	**6 666**	**114 524**
Afghanistan	6 389	0	0	6 389	1 742	0	2	1 740	400	0	0	400
Armenia	0	0	0	0	0	0	0	0	0	0	0	0
Azerbaijan	0	0	0	0	0	3	0	2	0	2	0	1
Bahrain	0	10	0	10	0	1	1	0	0	30	0	30
Bangladesh	32 020	0	0	32 020	612	10	0	621	70	0	0	71
Bhutan	1 381	68	0	1 449	45	0	0	45	18	0	0	18
Brunei Darussalam	79	2	0	81	216	0	0	217	90	35	0	125
Cambodia	6 907	0	47	6 860	1 040	0	459	581	175	0	155	20
China	204 239	185	215	204 209	108 718	7 169	3 480	112 407	26 969	2 684	753	28 901
China – Hong Kong SAR	206	93	11	288	0	816	336	480	441	636	427	650
Cyprus	46	16	0	61	35	5	0	40	16	77	4	89
DPR of Korea	4 369	0	0	4 369	600	42	405	237	280	3	11	272
Georgia	0	0	0	0	0	0	0	0	0	0	0	0
India	279 350	0	7	279 343	24 989	336	23	25 302	17 460	17	27	17 450
Indonesia	153 540	0	1 039	152 501	47 245	178	683	46 739	7 338	33	429	6 941
Iran, Islamic Republic of	2 573	0	0	2 573	4 902	7	16	4 894	159	0	0	160
Iraq	111	0	0	111	50	0	0	50	8	0	0	8
Israel	13	24	0	37	100	55	0	154	0	772	1	771
Japan	360	310	9	661	22 897	47 860	7	70 750	24 493	11 528	10	36 011
Jordan	8	0	0	8	4	36	0	40	0	129	0	129
Kazakhstan	315	0	0	315	0	0	1	-1	0	4	0	4
Kuwait	0	41	0	41	0	3	0	3	0	56	0	56
Kyrgyzstan	0	0	0	0	0	2	0	2	0	50	0	49
Laos	4 642	0	0	4 641	785	0	417	368	310	0	119	191
Lebanon	518	13	0	531	7	6	0	13	9	197	3	204
Macau	0	6	0	6	0	18	0	18	0	11	5	7
Malaysia	10 035	39	186	9 888	35 771	224	7 152	28 843	8 382	409	3 805	4 985
Maldives	0	1	0	1	0	0	0	0	0	0	0	0
Mongolia	376	0	0	376	508	0	4	503	129	1	115	16
Myanmar	20 612	0	58	20 554	3 023	0	498	2 525	351	0	172	179
Nepal	20 718	0	0	20 718	620	0	0	620	620	0	0	620
Oman	0	2	1	0	0	3	5	- 2	0	131	0	131
Pakistan	27 640	0	0	27 640	2 062	130	0	2 192	1 280	62	0	1 342
Philippines	37 280	0	211	37 069	3 394	636	13	4 017	313	567	145	735
Qatar	0	12	0	12	0	2	0	2	0	14	0	14
Republic of Korea	4 497	396	0	4 893	1 994	9 066	2	11 058	3 440	1 161	24	4 577
Saudi Arabia	0	37	0	36	0	40	0	40	0	602	7	595
Singapore	210	78	203	86	0	55	26	30	25	758	268	514
Sri Lanka	9 780	4	0	9 783	710	0	0	710	5	16	0	20
Syrian Arab Republic	20	0	0	20	35	5	0	40	9	126	0	135
Tajikistan	0	0	0	0	0	0	0	0	0	4	0	4
Thailand	36 894	154	53	36 995	2 818	939	388	3 370	325	2 296	45	2 575
Turkey	8 534	165	1	8 697	10 745	820	64	11 501	4 331	227	103	4 455
Turkmenistan	0	0	0	0	0	0	0	0	0	4	0	4
United Arab Emirates	0	26	1	25	0	32	0	32	0	273	0	273
Uzbekistan	0	0	0	0	0	1	0	1	0	1	0	1
Viet Nam	31 250	0	72	31 178	4 487	1	350	4 139	721	8	31	698
Yemen	324	0	0	324	0	5	0	5	0	99	0	98

Note: The regional breakdown reflects geographical rather than economic or political groupings.

Wood-based panels (thousand m³)				Pulp for paper (thousand m³)				Paper and paperboard (thousand m³)				Country
Production	Imports	Exports	Consumption	Production	Imports	Exports	Consumption	Production	Imports	Exports	Consumption	
5	0	0	5	0	0	0	0	3	2	0	5	Uganda
4	0	0	4	54	0	0	54	25	3	2	26	United Republic of Tanzania
18	0	0	18	0	0	0	0	2	5	0	7	Zambia
79	5	25	59	42	2	0	44	84	22	4	102	Zimbabwe
45 292	**17 637**	**15 741**	**47 188**	**42 869**	**9 847**	**1 383**	**51 333**	**81 926**	**15 009**	**7 177**	**89 758**	**ASIA**
1	0	0	2	0	0	0	0	0	1	0	1	Afghanistan
0	0	0	0	0	0	0	0	0	0	0	0	Armenia
0	1	0	0	0	0	0	0	0	0	0	0	Azerbaijan
0	16	0	16	0	11	0	11	0	11	0	11	Bahrain
9	1	0	10	122	6	0	128	160	103	0	263	Bangladesh
13	0	0	13	0	0	0	0	0	0	0	0	Bhutan
0	37	0	37	0	0	0	0	0	1	0	1	Brunei Darussalam
58	1	29	30	0	0	0	0	0	1	0	1	Cambodia
15 349	4 612	483	19 479	24 751	2 079	20	26 809	30 253	3 622	958	32 917	China
41	1 277	818	500	0	12	6	6	280	3 530	2 200	1 610	China – Hong Kong SAR
21	54	2	74	0	4	0	4	0	49	1	48	Cyprus
0	8	0	8	106	24	0	130	80	3	0	83	DPR of Korea
10	0	0	10	0	0	0	0	0	0	0	0	Georgia
348	20	20	348	1 870	265	3	2 132	3 025	350	6	3 369	India
10 128	47	8 302	1 873	2 635	637	1 129	2 143	4 386	199	1 213	3 372	Indonesia
383	1	0	384	245	24	0	269	205	362	0	567	Iran, Islamic Republic of
3	0	0	3	9	0	0	9	18	0	0	18	Iraq
177	293	39	431	0	99	0	99	275	350	15	609	Israel
7 048	6 704	34	13 718	11 065	3 391	63	14 394	30 014	1 544	964	30 595	Japan
0	56	0	56	8	39	0	47	31	106	6	131	Jordan
0	2	0	2	0	0	0	0	0	2	0	1	Kazakhstan
0	82	0	82	0	10	0	10	0	42	0	42	Kuwait
0	18	0	17	0	0	0	0	0	6	0	6	Kyrgyzstan
12	0	9	5	0	0	0	0	0	0	0	0	Laos
46	119	0	165	0	20	0	20	42	77	12	107	Lebanon
0	29	5	24	0	0	0	0	0	4	0	3	Macau
6 770	176	5 186	1 760	103	64	0	167	674	808	41	1 442	Malaysia
0	4	0	4	0	0	0	0	0	1	0	1	Maldives
2	1	0	3	0	0	0	0	0	2	0	1	Mongolia
25	2	19	7	9	0	0	9	15	13	0	28	Myanmar
0	0	0	0	15	0	0	15	13	0	0	13	Nepal
0	38	0	38	0	3	0	3	0	7	0	6	Oman
110	17	0	127	165	78	0	243	447	215	0	661	Pakistan
596	278	41	833	149	64	12	201	613	348	13	948	Philippines
0	14	0	14	0	0	0	0	0	2	0	2	Qatar
2 136	1 644	120	3 660	618	2 242	0	2 860	7 681	676	1 384	6 973	Republic of Korea
0	601	2	599	0	98	1	96	0	311	2	309	Saudi Arabia
355	657	325	687	0	31	16	15	87	891	210	768	Singapore
15	12	0	26	10	11	0	21	25	78	0	103	Sri Lanka
27	49	0	76	0	11	0	11	1	52	0	53	Syrian Arab Republic
0	0	0	0	0	0	0	0	0	0	0	0	Tajikistan
493	142	181	454	503	346	131	718	2 241	468	204	2 506	Thailand
1 078	315	77	1 316	354	146	0	500	1 235	574	37	1 772	Turkey
0	0	0	0	0	0	0	0	0	0	0	0	Turkmenistan
0	244	26	218	0	13	0	13	0	102	2	100	United Arab Emirates
0	2	0	2	0	0	0	0	0	5	0	5	Uzbekistan
39	39	20	58	133	35	0	167	125	71	2	194	Viet Nam
0	22	0	22	0	1	0	1	0	19	0	19	Yemen

Country	Fuelwood and charcoal (thousand m³)				Industrial roundwood (thousand m³)				Sawnwood (thousand m³)			
	Production	Imports	Exports	Consumption	Production	Imports	Exports	Consumption	Production	Imports	Exports	Consumption
OCEANIA	**8 756**	**0**	**12**	**8 745**	**41 461**	**14**	**18 347**	**23 128**	**6 764**	**861**	**1 067**	**6 557**
American Samoa	0	0	0	0	0	0	0	0	0	0	0	0
Australia	2 904	0	12	2 892	19 813	7	7 895	11 924	3 347	742	55	4 034
Cook Islands	0	0	0	0	5	0	4	0	0	3	0	3
Fiji	37	0	0	37	561	0	356	205	102	0	16	86
French Polynesia	0	0	0	0	0	0	0	0	0	34	0	34
Guam	0	0	0	0	0	0	0	0	0	0	0	0
Kiribati	0	0	0	0	0	0	0	0	0	2	0	2
Marshall Islands	0	0	0	0	0	0	0	0	0	6	0	6
Micronesia, Fed. States of	0	0	0	0	0	0	0	0	0	7	0	7
Nauru	0	0	0	0	0	0	0	0	0	0	0	0
New Caledonia	0	0	0	0	5	0	0	5	3	16	0	19
New Zealand	50	0	0	50	16 999	3	6 275	10 727	3 052	39	961	2 130
Niue	0	0	0	0	0	0	0	0	0	0	0	0
N. Mariana Islands	0	0	0	0	0	0	0	0	0	0	0	0
Palau	0	0	0	0	0	0	8	- 7	0	3	0	3
Papua New Guinea	5 533	0	0	5 533	3 239	0	3 039	201	218	0	17	201
Samoa	70	0	0	70	61	0	0	62	21	2	0	22
Solomon Islands	138	0	0	138	734	0	765	- 31	12	0	11	2
Tonga	0	0	0	0	5	0	0	5	1	2	0	4
Tuvalu	0	0	0	0	0	0	0	0	0	0	0	0
Vanuatu	24	0	0	24	39	0	4	36	7	0	6	1
EUROPE	**82 439**	**2 948**	**2 062**	**83 325**	**369 650**	**44 700**	**46 074**	**368 276**	**110 285**	**33 251**	**41 865**	**101 671**
Albania	346	0	0	346	64	0	1	63	5	0	0	4
Austria	3 059	346	1	3 404	11 346	5 708	2 236	14 818	7 804	1 030	4 100	4 734
Belarus	809	0	0	809	9 206	0	516	8 690	1 545	4	152	1 397
Belgium and Luxembourg	550	149	37	662	3 635	3 503	1 315	5 822	1 209	2 061	561	2 709
Bosnia and Herzegovina	0	0	0	0	40	0	0	40	20	0	1	19
Bulgaria	1 203	0	59	1 144	2 020	3	186	1 837	253	0	36	218
Croatia	1 067	0	0	1 067	1 845	139	284	1 700	598	59	408	248
Czech Republic	689	19	174	534	12 317	217	3 239	9 295	3 490	263	1 407	2 347
Denmark	491	69	8	552	1 797	716	413	2 100	583	1 921	94	2 410
Estonia	604	4	31	578	3 297	130	2 221	1 206	400	22	506	- 85
Finland	4 094	62	4	4 152	42 503	7 323	749	49 077	9 270	152	7 044	2 378
France	10 460	148	454	10 154	30 980	2 753	3 650	30 083	9 590	2 214	1 237	10 566
Germany	3 795	551	64	4 282	35 543	1 639	4 148	33 034	14 335	4 817	1 871	17 281
Greece	1 350	4	4	1 350	674	28	5	697	337	466	10	793
Hungary	2 032	9	303	1 738	1 894	243	657	1 480	241	825	166	900
Iceland	0	3	0	3	0	6	0	6	0	93	0	92
Ireland	66	3	0	69	2 225	7	515	1 717	687	286	269	704
Italy	4 970	436	4	5 402	4 163	5 398	9	9 552	1 650	5 527	99	7 078
Latvia	1 217	0	67	1 150	5 690	7	2 310	3 387	1 300	3	1 407	- 105
Liechtenstein	0	0	0	0	9	0	0	9	0	0	0	0
Lithuania	1 220	0	4	1 217	4 310	18	963	3 365	1 250	30	1 076	204
Malta	0	1	0	1	0	7	2	5	0	6	0	6
Netherlands	160	159	53	266	828	676	605	899	367	3 398	389	3 376
Norway	470	317	4	783	7 695	3 123	530	10 288	2 405	799	709	2 495
Poland	1 688	15	205	1 498	17 783	486	484	17 785	5 060	150	1 020	4 190
Portugal	598	17	23	593	8 850	1 530	494	9 886	1 731	222	717	1 236
Republic of Moldova	0	0	0	0	484	28	0	512	25	57	0	82
Romania	3 175	0	72	3 103	9 441	75	0	9 516	1 693	2	887	808
Russian Federation	29 250	0	79	29 171	67 000	964	16 312	51 652	21 600	50	4 723	16 927
Slovakia	454	0	47	407	4 887	2	545	4 343	629	8	352	285
Slovenia	235	18	76	177	1 709	331	116	1 924	513	117	382	249

Note: The regional breakdown reflects geographical rather than economic or political groupings.

Wood-based panels (thousand m³)				Pulp for paper (thousand m³)				Paper and paperboard (thousand m³)				Country
Production	Imports	Exports	Consumption	Production	Imports	Exports	Consumption	Production	Imports	Exports	Consumption	
2 158	**360**	**813**	**1 704**	**2 330**	**201**	**680**	**1 851**	**2 651**	**1 307**	**625**	**3 333**	**OCEANIA**
0	0	0	0	0	0	0	0	0	0	0	0	American Samoa
1 085	199	206	1 078	975	184	0	1 159	1 850	1 111	241	2 719	Australia
0	0	0	0	0	0	0	0	0	0	0	0	Cook Islands
16	120	19	117	0	0	0	0	0	14	0	14	Fiji
0	10	0	10	0	0	0	0	0	2	0	2	French Polynesia
0	3	0	3	0	0	0	0	0	0	0	0	Guam
0	0	0	0	0	0	0	0	0	0	0	0	Kiribati
0	3	0	3	0	0	0	0	0	0	0	0	Marshall Islands
0	0	0	0	0	0	0	0	0	0	0	0	Micronesia, Fed. States of
0	0	0	0	0	0	0	0	0	0	0	0	Nauru
0	5	0	5	0	0	0	0	0	5	0	5	New Caledonia
1 042	12	588	466	1 355	16	680	692	801	169	384	586	New Zealand
0	0	0	0	0	0	0	0	0	0	0	0	Niue
0	0	0	0	0	0	0	0	0	0	0	0	N. Mariana Islands
0	1	0	1	0	0	0	0	0	0	0	0	Palau
15	2	0	16	0	0	0	0	0	6	0	5	Papua New Guinea
0	1	0	1	0	0	0	0	0	0	0	0	Samoa
0	0	0	0	0	0	0	0	0	0	0	0	Solomon Islands
0	0	0	0	0	0	0	0	0	0	0	0	Tonga
0	0	0	0	0	0	0	0	0	0	0	0	Tuvalu
0	0	0	0	0	0	0	0	0	0	0	0	Vanuatu
44 167	**15 270**	**15 759**	**43 678**	**38 047**	**14 283**	**9 290**	**43 040**	**80 936**	**33 213**	**41 236**	**72 913**	**EUROPE**
16	1	0	17	16	6	0	22	44	2	0	46	Albania
1 961	348	1 274	1 035	1 396	575	287	1 684	3 653	881	3 030	1 505	Austria
374	26	52	348	31	5	0	36	131	9	7	133	Belarus
2 609	655	2 279	985	382	534	193	723	1 432	2 071	1 015	2 489	Belgium and Luxembourg
34	0	0	34	0	0	0	0	0	0	0	0	Bosnia and Herzegovina
233	8	13	228	107	2	32	77	148	47	6	189	Bulgaria
73	62	23	112	99	9	36	72	304	179	95	388	Croatia
975	151	454	672	490	99	225	364	616	273	249	641	Czech Republic
432	570	116	886	180	52	69	162	345	1 044	242	1 148	Denmark
313	206	312	207	28	0	0	28	21	36	8	49	Estonia
1 538	147	1 154	531	9 676	57	1 535	8 198	10 441	186	8 529	2 098	Finland
4 150	1 221	1 954	3 416	2 516	2 037	359	4 194	8 531	3 097	2 523	9 105	France
10 428	3 700	2 088	12 040	1 816	3 456	332	4 940	14 733	6 770	6 319	15 184	Germany
363	151	34	481	27	70	0	97	750	251	25	976	Greece
459	65	187	337	21	29	19	31	363	288	129	522	Hungary
0	24	0	24	0	0	0	0	0	26	0	25	Iceland
434	75	212	297	0	8	0	8	0	675	67	608	Ireland
3 922	819	504	4 237	563	2 866	20	3 409	6 954	3 023	1 790	8 187	Italy
236	6	153	90	9	2	0	11	6	42	7	42	Latvia
0	0	0	0	0	0	0	0	0	0	0	0	Liechtenstein
203	53	143	114	9	6	3	12	33	38	25	46	Lithuania
0	13	0	13	0	0	0	0	0	19	0	19	Malta
99	1 398	228	1 270	128	1 024	265	887	2 987	2 142	1 915	3 214	Netherlands
535	170	237	468	2 269	103	390	1 983	2 096	519	1 832	783	Norway
2 410	331	493	2 248	848	188	67	969	1 525	582	418	1 689	Poland
1 170	115	647	638	1 608	89	1 000	697	1 026	491	630	887	Portugal
10	20	0	29	0	0	0	0	0	18	5	13	Republic of Moldova
433	104	102	435	247	15	4	258	332	66	78	320	Romania
3 036	29	939	2 126	3 725	12	979	2 758	3 212	56	1 480	1 788	Russian Federation
312	44	56	300	235	19	51	203	327	81	218	190	Slovakia
361	92	150	303	101	139	49	192	460	85	177	368	Slovenia

Country	Fuelwood and charcoal (thousand m³)				Industrial roundwood (thousand m³)				Sawnwood (thousand m³)			
	Production	Imports	Exports	Consumption	Production	Imports	Exports	Consumption	Production	Imports	Exports	Consumption
Spain	2 707	116	158	2 665	12 997	2 113	266	14 844	3 262	1 591	50	4 804
Sweden	3 824	113	8	3 929	52 600	5 737	1 931	56 406	14 370	214	11 665	2 919
Switzerland	930	65	109	886	3 370	861	1 156	3 075	1 355	518	127	1 746
The FYR of Macedonia	616	31	0	647	158	0	2	157	40	27	63	3
Ukraine	0	86	8	78	0	3	131	- 128	0	19	146	- 126
United Kingdom	261	204	7	458	7 021	772	68	7 725	2 258	5 952	62	8 148
Yugoslavia	50	0	0	51	1 270	154	13	1 410	410	335	129	616
NORTH AND CENTRAL AMERICA	**155 437**	**417**	**684**	**155 170**	**600 447**	**9 433**	**23 701**	**586 180**	**176 948**	**47 096**	**58 185**	**165 858**
Antigua and Barbuda	0	0	0	0	0	0	0	0	0	11	0	11
Bahamas	0	0	0	0	117	14	0	131	1	87	0	87
Barbados	0	0	0	0	5	1	5	1	0	34	0	34
Belize	126	0	0	126	62	0	4	58	20	0	4	17
Bermuda	0	0	0	0	0	12	2	9	0	7	0	7
British Virgin Islands	0	0	0	0	0	0	0	0	0	4	0	4
Canada	5 319	104	234	5 189	183 113	7 022	2 607	187 528	62 829	1 697	50 458	14 068
Cayman Islands	0	0	0	0	0	2	0	2	0	14	0	14
Costa Rica	3 440	0	0	3 440	1 651	8	4	1 655	780	3	4	779
Cuba	2 541	0	0	2 541	611	0	0	612	130	15	1	144
Dominica	0	0	0	0	0	0	0	0	0	8	0	8
Dominican Republic	976	7	0	983	6	0	0	7	0	234	0	234
El Salvador	6 809	0	0	6 809	211	2	0	213	70	72	0	142
Greenland	0	0	0	0	0	0	0	0	0	7	0	7
Grenada	0	0	0	0	0	3	0	3	0	12	0	12
Guadeloupe	15	0	0	16	0	6	0	6	1	69	0	70
Guatemala	13 328	0	0	13 328	795	2	0	797	355	2	47	310
Haiti	6 305	3	0	6 308	239	4	0	243	14	21	0	34
Honduras	6 038	0	0	6 037	664	0	65	600	326	5	21	310
Jamaica	312	0	0	312	43	18	0	61	12	114	0	125
Martinique	10	0	0	11	2	3	0	5	1	29	0	30
Mexico	16 731	2	136	16 597	5 914	27	28	5 914	2 543	690	417	2 816
Montserrat	0	0	0	0	0	0	0	0	0	4	0	4
Netherlands Antilles	0	0	0	0	0	2	0	1	0	45	0	45
Nicaragua	3 786	1	0	3 787	267	3	2	267	155	0	91	64
Panama	969	0	0	970	118	24	6	136	37	4	1	40
Puerto Rico	n.a.	n.a.	n.a.	0	n.a.	n.a.	n.a.	0	n.a.	n.a.	n.a.	0
Saint Kitts and Nevis	0	0	0	0	0	1	0	1	0	5	0	5
Saint Lucia	0	0	0	0	0	7	0	7	0	15	0	15
Saint Pierre and Miquelon	0	0	0	0	0	0	0	0	0	2	0	2
Saint Vincent and Grenadines	0	0	0	0	0	0	0	0	0	12	0	12
Trinidad and Tobago	22	0	0	22	34	0	0	35	19	32	0	50
United States of America	88 710	295	313	88 692	406 595	2 266	20 977	387 884	109 654	43 823	7 139	146 338
United States Virgin Islands	0	0	0	0	0	0	0	0	0	0	0	0
SOUTH AMERICA	**193 007**	**63**	**420**	**192 651**	**129 890**	**40**	**10 066**	**119 863**	**28 336**	**371**	**3 493**	**25 214**
Argentina	4 498	1	332	4 167	6 220	6	968	5 258	1 000	171	50	1 121
Bolivia	1 419	0	0	1 418	892	0	7	885	176	0	143	33
Brazil	135 652	46	63	135 635	84 711	8	2 215	82 504	19 091	112	1 640	17 563
Chile	9 984	1	0	9 985	21 387	2	6 451	14 938	3 802	24	1 087	2 739
Colombia	18 062	1	0	18 063	2 703	2	8	2 697	644	12	10	646
Ecuador	5 474	0	4	5 470	5 514	0	20	5 495	1 886	0	83	1 803
French Guiana	72	0	0	72	60	0	2	59	15	0	4	12
Guyana	21	0	1	20	468	0	22	446	101	0	19	82

Note: The regional breakdown reflects geographical rather than economic or political groupings.

Wood-based panels *(thousand m³)*				Pulp for paper *(thousand m³)*				Paper and paperboard *(thousand m³)*				Country
Production	Imports	Exports	Consumption	Production	Imports	Exports	Consumption	Production	Imports	Exports	Consumption	
2 735	765	645	2 856	1 462	493	655	1 300	3 767	2 175	515	5 427	Spain
971	614	388	1 197	9 778	239	2 609	7 408	9 038	444	7 483	1 999	Sweden
667	479	467	679	244	426	94	576	1 461	1 064	1 044	1 481	Switzerland
0	9	6	3	1	3	0	4	12	23	0	35	The FYR of Macedonia
0	19	15	4	0	0	0	0	0	26	9	17	Ukraine
2 543	2 736	429	4 850	575	1 701	16	2 260	6 189	6 385	1 366	11 208	United Kingdom
132	44	7	168	0	0	0	0	0	94	2	92	Yugoslavia
49 911	**9 603**	**9 974**	**49 539**	**83 448**	**5 799**	**15 567**	**73 680**	**106 847**	**14 997**	**22 844**	**99 000**	**NORTH AND CENTRAL AMERICA**
0	4	0	4	0	0	0	0	0	0	0	0	Antigua and Barbuda
0	15	0	15	0	0	0	0	0	4	0	4	Bahamas
0	7	0	7	0	0	0	0	0	6	0	6	Barbados
0	3	0	3	0	0	0	0	0	2	0	2	Belize
0	2	0	2	0	0	0	0	0	5	0	5	Bermuda
0	1	0	1	0	0	0	0	0	0	0	0	British Virgin Islands
9 966	1 096	6 860	4 202	24 403	267	10 133	14 537	18 414	1 584	13 393	6 605	Canada
0	5	0	5	0	0	0	0	0	0	0	0	Cayman Islands
74	13	29	58	10	3	0	13	20	258	4	274	Costa Rica
149	7	0	155	52	3	0	55	57	7	0	64	Cuba
0	3	0	3	0	0	0	0	0	4	0	4	Dominica
0	38	0	38	0	1	0	1	21	95	0	116	Dominican Republic
0	16	0	16	0	23	0	23	56	61	6	111	El Salvador
0	5	0	5	0	0	0	0	0	0	0	0	Greenland
0	3	0	3	0	0	0	0	0	0	0	0	Grenada
0	22	0	22	0	0	0	0	0	6	0	6	Guadeloupe
43	14	10	47	0	3	0	3	31	131	2	160	Guatemala
0	7	0	7	0	0	0	0	0	5	0	5	Haiti
14	8	2	19	0	0	0	0	0	92	0	91	Honduras
0	21	0	21	0	0	0	0	0	49	0	49	Jamaica
0	7	0	7	0	0	0	0	0	5	0	5	Martinique
606	270	112	764	511	398	23	887	3 047	752	305	3 494	Mexico
0	0	0	0	0	0	0	0	0	0	0	0	Montserrat
0	18	0	18	0	2	0	2	0	4	0	4	Netherlands Antilles
5	5	0	10	0	0	0	0	0	6	0	6	Nicaragua
21	9	0	30	0	1	0	1	28	112	19	120	Panama
n.a.	n.a.	n.a.	0	n.a.	n.a.	n.a.	0	n.a.	n.a.	n.a.	0	Puerto Rico
0	1	0	1	0	0	0	0	0	0	0	0	Saint Kitts and Nevis
0	7	0	7	0	0	0	0	0	10	0	10	Saint Lucia
0	0	0	0	0	0	0	0	0	0	0	0	Saint Pierre and Miquelon
0	3	0	3	0	0	0	0	0	4	0	4	Saint Vincent and Grenadines
0	21	0	21	0	3	0	3	0	38	0	38	Trinidad and Tobago
39 033	7 963	2 960	44 037	58 472	5 092	5 411	58 153	85 173	11 752	9 113	87 812	United States of America
0	0	0	0	0	0	0	0	0	0	0	0	United States Virgin Islands
6 037	**409**	**2 098**	**4 348**	**9 719**	**673**	**4 023**	**6 369**	**9 328**	**2 644**	**1 519**	**10 453**	**SOUTH AMERICA**
590	68	210	448	822	136	190	768	1 108	548	26	1 630	Argentina
2	1	0	3	0	8	0	8	2	52	0	53	Bolivia
3 558	223	1 164	2 617	6 225	209	2 176	4 258	5 885	909	1 220	5 574	Brazil
844	20	418	446	2 123	11	1 655	479	597	151	142	606	Chile
176	35	10	201	307	73	0	380	676	348	63	961	Colombia
380	6	90	296	0	5	0	5	86	319	7	399	Ecuador
0	3	0	3	0	0	0	0	0	0	0	0	French Guiana
100	0	97	3	0	0	0	0	0	2	0	2	Guyana

Country	Fuelwood and charcoal (thousand m³)				Industrial roundwood (thousand m³)				Sawnwood (thousand m³)			
	Production	Imports	Exports	Consumption	Production	Imports	Exports	Consumption	Production	Imports	Exports	Consumption
Paraguay	6 524	0	18	6 506	3 877	6	1	3 882	400	0	386	14
Peru	7 315	10	0	7 325	1 546	4	6	1 545	693	2	13	682
Suriname	19	0	0	19	103	0	21	82	29	0	3	26
Uruguay	3 050	4	0	3 054	1 043	2	298	747	269	43	54	258
Venezuela	918	0	0	917	1 366	7	48	1 326	230	5	2	234
WORLD TOTAL	**1 864 760**	**5 509**	**5 842**	**1 864 427**	**1 489 530**	**123 372**	**119 963**	**1 492 939**	**429 645**	**109 200**	**112 612**	**426 233**

Note: The regional breakdown reflects geographical rather than economic or political groupings.

Wood-based panels (thousand m³)				Pulp for paper (thousand m³)				Paper and paperboard (thousand m³)				Country
Production	Imports	Exports	Consumption	Production	Imports	Exports	Consumption	Production	Imports	Exports	Consumption	
96	2	98	0	0	0	0	0	13	26	1	38	Paraguay
83	14	5	92	48	22	0	70	140	126	4	262	Peru
7	0	3	5	0	0	0	0	0	3	0	3	Suriname
6	17	0	22	29	15	2	42	86	47	8	126	Uruguay
194	20	2	212	165	193	0	358	735	114	50	799	Venezuela
149 385	**43 822**	**44 732**	**148 474**	**178 543**	**31 115**	**31 696**	**177 962**	**284 383**	**68 375**	**73 990**	**278 767**	**WORLD TOTAL**

TABLE 5

International conventions and agreements and national forest programmes

Country	Status of ratification of UNCED conventions			Members under ITTA 1994	Countries with NFPs	
	CBD	FCCC	UNCCD		Programme starting date	Programme type
AFRICA						
Algeria	x	x	x			
Angola	x		x		1994	FSR
Benin	x	x	x		1993	NFAP
Botswana	x	x	x			
Burkina Faso	x	x	x		1986/91/92	NPCD/NFAP/NEAP
Burundi	x	x	x		1990	NFAP/NEAP
Cameroon	x	x	x	x	1986	NFAP
Cape Verde	x	x	x		1992	NFAP
Central African Republic	x	x	x	x	1990	NFAP
Chad	x	x	x		1989	NPCD
Comoros	x	x	x			
Congo	x				1987	NFAP/NEAP
Congo, Democratic Republic of	x	x	x	x	1987	NFAP
Côte d'Ivoire	x	x	x	x		
Djibouti	x	x	x			
Egypt	x	x	x	x		
Equatorial Guinea	x		x	x	1989	NFAP
Eritrea	x	x	x		1997	NEAP
Ethiopia	x	x	x		1990	NFAP
Gabon	x	x	x	x	1994	NFAP
Gambia	x	x	x		1991	NFAP
Ghana	x	x	x	x	1986/97	NFAP/MP
Guinea	x	x	x		1986/89	NFAP/PNAE
Guinea-Bissau	x	x	x		1990/95	MP/NP
Kenya	x	x	x		1991	MP
Lesotho	x	x	x			
Liberia			x	x	1989	NFAP*
Libyan Arab Jamahiriya			x			
Madagascar	x		x		1989/94/95	NFAP/NEAP/MP
Malawi	x	x	x			
Mali	x	x	x		1987/95	NFAP/NPCD/NEAP
Mauritania	x	x	x			
Mauritius	x	x	x			
Morocco	x	x	x			
Mozambique	x	x	x		1989	NP*
Namibia	x	x	x			
Niger	x	x	x		1985/90	NPCD/NFAP
Nigeria	x	x	x		1990	NFAP
Réunion						
Rwanda	x				1987/90	NP/NFAP
Saint Helena						
Sao Tome and Principe			x			
Senegal	x	x	x		1988/90	NPCD/NFAP
Seychelles	x	x	x			
Sierra Leone	x	x	x		1988	FSR/NFAP
Somalia						
South Africa	x	x	x		1989	NFAP
Sudan	x	x	x		1984/86	NFAP/FSR
Swaziland	x	x	x			
Togo	x	x	x	x	1990	NFAP
Tunisia	x	x	x			
Uganda	x	x	x		1994	NFAP

Note: The regional breakdown reflects geographical rather than economic or political groupings.

Country	Status of ratification of UNCED conventions			Members under ITTA 1994	Countries with NFPs	
	CBD	FCCC	UNCCD		Programme starting date	Programme type
United Republic of Tanzania	x	x	x		1989	NFAP
Zambia	x	x	x		1992	NFAP
Zimbabwe	x	x	x		1998	NFAP
ASIA						
Afghanistan			x			
Armenia	x	x	x		1994	NP
Azerbaijan		x	x			
Bahrain	x	x	x			
Bangladesh	x	x	x		1989	MP
Bhutan	x	x			1989	MP
Brunei Darussalam						
Cambodia	x	x	x	x	1994	NFAP
China	x	x	x	x	1992	NFAP
China – Hong Kong SAR						
Cyprus	x	x			1991	NP
DPR of Korea	x	x				
Georgia	x	x				
India	x	x	x	x	1992/93/93	NCS/NEAP/NFAP
Indonesia	x	x	x	x	1987	NFAP
Iran, Islamic Republic of	x	x	x			
Iraq						
Israel	x	x	x			
Japan	x	x		x	1996	NP
Jordan	x	x	x			
Kazakhstan	x	x	x			
Kuwait		x	x			
Kyrgyzstan	x		x			
Laos	x	x	x		1989	NFAP
Lebanon	x	x	x		1994	NP
Macau						
Malaysia	x	x	x	x	1984-96	NP
Maldives	x	x				
Mongolia	x	x	x		1991/93/97	NPCD/NEAP/NFAP
Myanmar	x	x	x	x	1995	NFAP
Nepal	x	x	x	x	1984	MP
Oman	x	x	x			
Pakistan	x	x	x		1992	MP
Philippines	x	x		x	1988	MP
Qatar	x	x				
Republic of Korea	x	x		x	1998	NP
Saudi Arabia		x	x			
Singapore	x	x				
Sri Lanka	x	x			1983	MP
Syrian Arab Republic	x	x	x			
Tajikistan	x	x	x			
Thailand		x		x	1991	MP
Turkey	x		x		1990	MP
Turkmenistan	x	x	x			
United Arab Emirates		x				
Uzbekistan	x	x	x			
Viet Nam	x	x	x		1988	NFAP
Yemen	x	x	x			

Note: The regional breakdown reflects geographical rather than economic or political groupings.

Country	Status of ratification of UNCED conventions			Members under ITTA 1994	Countries with NFPs	
	CBD	FCCC	UNCCD		Programme starting date	Programme type
OCEANIA						
American Samoa						
Australia	x	x		x	1992/95	NP
Cook Islands	x	x				
Fiji	x	x	x	x	1989	NFAP
French Polynesia						
Guam						
Kiribati	x	x				
Marshall Islands	x	x	x			
Micronesia, Fed. States of	x	x	x		1991/94	NEMS/NFAP
Nauru	x	x				
New Caledonia						
New Zealand	x	x		x	1987	NP
Niue	x	x	x			
N. Mariana Islands						
Palau						
Papua New Guinea	x	x		x	1989	NP
Samoa	x	x			1988	NEMS
Solomon Islands	x	x			1991/94	NEMS/NP
Tonga	x				1992	NEMS
Tuvalu		x				
Vanuatu	x	x			1991	NFAP
EUROPE						
Albania	x	x				
Austria	x	x	x	x	1996	NP
Belarus	x				1997	NP
Belgium and Luxembourg	x	x	x	x	1990	NP
Bosnia and Herzegovina						
Bulgaria	x	x			1997	NP
Croatia	x	x			1990	NP
Czech Republic	x	x			1996	NP
Denmark	x	x	x	x	1994	NP
Estonia	x	x			1997	NP
Finland	x	x	x	x	1998	NP
France	x	x	x	x		
Germany	x	x	x	x	1996	NP
Greece	x	x	x	x		
Hungary	x	x			1996	NP
Iceland	x	x	x			
Ireland	x	x	x	x	1996	NP
Italy	x	x	x	x	1988	NP
Latvia	x	x			1994	NP
Liechtenstein	x	x			1991	NP
Lithuania	x	x			1996	NP
Malta			x			
Monaco	x	x				
Netherlands	x	x	x	x	1993	NP
Norway	x	x	x	x	1993	NP
Poland	x	x			1997	NP
Portugal	x	x	x	x	1996	NP
Republic of Moldova	x	x				
Romania	x	x				
Russian Federation	x	x			1997	NP
San Marino	x	x				
Slovakia	x	x			1994	NP
Slovenia	x	x			1993	NP

Note: The regional breakdown reflects geographical rather than economic or political groupings.

Country	Status of ratification of UNCED conventions			Members under ITTA 1994	Countries with NFPs	
	CBD	FCCC	UNCCD		Programme starting date	Programme type
Spain	x	x	x	x		
Sweden	x	x	x	x	1994	NP
Switzerland	x	x	x	x	1993	NP
The FYR of Macedonia	x	x				
Ukraine	x	x			1994	NP
United Kingdom	x	x	x	x	1994	NP
Yugoslavia		x				
NORTH AND CENTRAL AMERICA						
Antigua and Barbuda	x	x	x		1991	NFAP
Bahamas	x	x			1997	NP
Barbados	x	x	x		1993	NFAP
Belize	x	x	x		1988	NFAP
Bermuda						
British Virgin Islands						
Canada	x		x	x	1998	NP
Cayman Islands						
Costa Rica	x	x	x		1987	NFAP
Cuba	x	x	x		1992	NFAP
Dominica	x	x	x		1989	NEAP
Dominican Republic	x		x		1986	NFAP
El Salvador	x	x	x		1991	NFAP
Greenland						
Grenada	x	x	x		1992/93	NEAP/NAFP
Guadeloupe						
Guatemala	x	x			1989	NFAP
Haiti	x	x	x		1990	NFAP
Honduras	x	x	x	x	1986	NP
Jamaica	x	x	x		1988	NFAP
Martinique						
Mexico	x	x	x		1993	NP
Montserrat						
Netherlands Antilles						
Nicaragua	x	x	x		1991	NFAP
Panama	x	x	x	x	1988	NFAP
Puerto Rico						
Saint Kitts and Nevis	x	x	x		1988	NFAP
Saint Lucia	x	x	x		1990	NFAP
Saint Pierre and Miquelon						
Saint Vincent and Grenadines	x	x	x		1988	NFAP
Trinidad and Tobago	x	x			1990	NFAP
United States of America		x		x	1998	NP
United States Virgin Islands						
SOUTH AMERICA						
Argentina	x		x		1997	NP
Bolivia	x		x	x	1988	NFAP
Brazil	x	x	x	x	1994	NP
Chile	x	x	x		1990	NFAP
Colombia	x	x		x	1997	NFAP
Ecuador	x	x	x	x	1988	NFAP
French Guiana						
Guyana	x	x	x	x	1987	NFAP
Paraguay	x	x	x		1992	NP
Peru	x	x	x	x	1986	NFAP
Suriname	x	x		x	1989	NFAP
Uruguay	x	x			1994	NP
Venezuela	x	x	x	x	1989	NFAP*

Note: The regional breakdown reflects geographical rather than economic or political groupings.

BIBLIOGRAPHY

Barthlott, W., Lauer, W. & Placke, A. 1996. Global distribution of species diversity on vascular plants: towards a world map of phytodiversity. *Erdkunde*, 50: 317-327.

Bibby, C.J., Collar, N.J., Crosby, M.J., Heath, M.F., Imboden, C., Johnson, T.H., Long, A.J., Satterfield, A.J. & Thirgood, S.J. 1992. *Putting biodiversity on the map: priority areas for global conservation.* Cambridge, UK, ICBP.

Brockmann, K.L., Hemmelskamp, J. & Hohmeyer, O. 1996. *Certified tropical timber and consumer behaviour.* Mannheim, Germany, Centre for European Economic Research.

Brown, S., Sathaye, J., Cannell, M. & Kauppi, P. 1996. Management of forests for mitigation of greenhouse gas emissions. *In* R.T. Watson, M.C. Zinyowera & R.H. Moss, eds. *Climate change 1995 – impacts, adaptations and mitigation of climate change: scientific-technical analyses.* Contribution of Working Group II to the Second Assessment Report of the Intergovernmental Panel on Climate Change. Cambridge, UK, Cambridge University Press.

Bureau of Resource Sciences. 1997. *National forest inventory (1997): national plantation inventory of Australia.* Canberra, Australia.

Chipeta, M.E. 1996. *Forestry funding in Asia-Pacific, Africa and Latin America/Caribbean regions: perceptions of main opportunities and constraints.* Information Note to the UNDP/Denmark/South Africa Workshop on Financial Mechanisms and Sources of Finance for Sustainable Forestry, Pretoria, South Africa, June 1996. New York, NY, USA, UNDP.

Commission on Sustainable Development (CSD). 1997. *Report of the ad hoc Intergovernmental Panel on Forests on its fourth session, 20 March 1997.* Document of the fifth session of CSD, 7-25 April 1997. E/CN.17/1997/12. New York, NY, USA.

Department of Agriculture of Northern Ireland (DANI), Forestry Commission. 1998. *The UK Forestry Standard. The Government's approach to sustainable forestry.* Edinburgh, UK, Forestry Commission for Great Britain and Department of Agriculture for Northern Ireland.

Dudley, N., Gilmour, D. & Jeanrenaud, J.P. 1996. *Forests for life – the WWF/IUCN forest policy book.* Godalming, UK, WWF.

European Commission (EC) General Secretariat. 1997. *Energy for the future: Renewable sources of energy – A White Paper for a community strategy and action plan.* COM(97) 599/7.

FAO. 1982. *Tropical forest resources.* FAO Forestry Paper No. 30. Rome.

FAO. 1995. *Forest resources assessment 1990 – global synthesis.* FAO Forestry Paper No. 124. Rome.

FAO. 1996a. *FRA 1990: survey of tropical forest cover and study of change processes.* FAO Forestry Paper No. 130. Rome.

FAO. 1996b. *FAO model code of forest harvesting practice*, by D.P. Dykstra & R. Heinrich. Rome.

FAO. 1996c. *Basic principles and operational guidelines. Formulation, execution and revision of national forest programmes.* FAO, Rome.

FAO. 1996d. *Directory of forestry education and training institutions.* Rome.

FAO. 1996e. *Memoria – Reunión regional sobre generación de electricidad a partir de biomasa.* Montevideo, Uruguay, 23-27 October 1995. Serie Forestal No. 7. Santiago, Chile.

FAO. 1997a. *Implications of sustainable forest management for global fibre supply*, by J. Williams, P.N. Duinker & G. Bull. Working Paper GFSS/WP/03. Rome.

FAO. 1997b. *Forest harvesting in natural forests of the Republic of the Congo*, by R. Scharpenberg. Forest Harvesting Case-Study No. 7. Rome.

FAO. 1997c. *Environmentally sound forest harvesting: testing the applicability of the FAO Model Code in the Amazon in Brazil*, by N. Winkler. Forest Harvesting Case-Study No. 8. Rome.

FAO. 1997d. *State of the World's Forests 1997.* Rome.

FAO. 1997e. *In-depth country study – New Zealand. Asia-Pacific forestry sector outlook study*, by C. Brown. Working Paper APFSOS/WP/05. Rome.

FAO. 1997f. *Aménagement des forêts naturelles des zones tropicales sèches*, by R. Bellefontaine, R. Gaston & Y. Petrucci. FAO Conservation Guide No. 32. Rome.

FAO. 1997g. *Regional study – the South Pacific.* Asia-Pacific Forestry Sector Outlook Study. Working Paper APFSOS/WP/01. Rome.

FAO. 1997h. *The role of wood energy in Europe and OECD. Wood energy today for tomorrow*, by R. van den Broek. Rome.

FAO. 1997i. *La generación de electricidad a partir de eucalipto y bagazo en ingenios azucareros de Nicaragua. Informe final*, by R. van den Broek. Rome.

FAO. 1997j. *Forest resources and roundwood supply in the Asia-Pacific countries: situation and outlook to the year 2010*, by J. Blanchez. APFSOS Working Paper No. 17. Rome.

FAO. 1997k. *New institutional trends in forestry research – case studies from Africa.* Rome.

FAO. 1997l. *Update on sustainable forest management and certification.* Advisory Committee on Paper and Wood Products, 38th session, Rome, 23-25 April 1997. ACPWP 97/5. Rome (http://www.fao.org/unfao/bodies/acpwp/acpwp38/sfmabse.htm)

FAO. 1998a. *Global Fibre Supply Model.* Rome.

FAO. 1998b. *Environmentally sound road construction in mountainous terrain: applying advanced operating methods and tools,* by N. Winkler. Forest Harvesting Case-Study No. 10. Rome.

FAO. 1998c. *Reduced impact timber harvesting in the tropical natural forest in Indonesia,* by Elias. Forest Harvesting Case-Study No. 11. Rome.

FAO. 1998d. *Environmentally sound forest infrastructure development and harvesting by long distance cable systems in the Himalayan region in Bhutan,* by N. Winkler. Forest Harvesting Case-Study No. 12. Rome.

FAO. 1998e. *Trade restrictions and their impacts on international trade in forest products,* by I.J. Bourke & J. Leitch. Rome.

FAO. 1998f. *Asia-Pacific forestry – towards 2010: report of the Asia-Pacific Forestry Sector Outlook Study.* Rome.

FAO. 1998g. *FAO Yearbook of Forest Products 1996.* Rome.

FAO. 1998h. *FAO Production Yearbook 1996.* Rome.

FAO. 1999. *Global Forest Products Outlook Study,* by A. Whiteman. Rome. (In press).

FAO/FORSPA. 1998. *Emerging institutional arrangements for forestry research.*

Houghton, J.T., Meira Filho, L.G., Callander, B.A., Harris, N., Kattenberg, A. & Maskell, K., eds. 1996. *Climate change 1995 – the science of climate change.* Contribution of Working Group I to the Second Assessment Report of the Intergovernmental Panel on Climate Change. Cambridge, UK, Cambridge University Press.

Intergovernmental Panel on Climate Change (IPCC). 1996. *Revised 1996 IPCC guidelines for national greenhouse gas inventories.* Vol. 1, *Greenhouse gas inventory reporting instructions.* Vol. 2, *Greenhouse gas inventory workbook.* Vol. 3, *Greenhouse gas inventory reference manual.* Geneva, Switzerland, IPCC/Paris, France, OECD, International Energy Agency (IEA).

Intergovernmental Panel on Forests (IPF). 1996. *Report of the Secretary-General on Programme Element I.4: Fragile ecosystems affected by desertification and the impact of air-borne pollution on forests.* E/CN.17/IPF/1996/3. Document prepared for the 2nd session of IPF. New York, NY, USA.

International Tropical Timber Organization (ITTO). 1998. *Tropical Timber Market Report,* 16-31 May.

Iremonger, S.I., Ravilious, C. & Quinton, T., eds. 1997. *A global overview of forest conservation.* CD-ROM. Cambridge, UK, WCMC and CIFOR.

Irland, L.C. 1998. Ice storms 1998 and the forests of the Northeast. *J. Forestry,* 96(9): 32-40.

Joint ECE/FAO Agriculture and Timber Division. 1996. *European timber trends and prospects into the 21st century.* Geneva Timber and Forest Study Papers (ECE) No. 11. New York, NY, USA, UN.

Kowero, G.S. & Spilsbury, M.J. 1997. *Capacity for forestry research in the Southern African Development Community.* Bogor, Indonesia, CIFOR.

Leach, G. 1993. *Farm trees and wood markets: a review and economic appraisal.* Pakistan Household Energy Strategy Study, UNDP/World Bank Energy Sector Management Assistance Programme. Washington, DC, USA, World Bank/Islamabad, Pakistan, Energy Wing, Government of Pakistan.

Lininger, H., Weingartner, R. & Grosjean, M. 1998. *Mountains of the world: water towers for the 21st century.* Bern, Switzerland, Mountain Agenda.

Markopoulos, M. 1998. *The impacts of certification on community forest enterprises: a case study of the Lomerio Community Forest Management Project, Bolivia.* London, UK, International Institute for Environment and Development.

Messerli, B. & Ives, J.D., eds. 1997. *Mountains of the world. A global priority.* New York, NY, USA, Parthenon Publishing Group, Inc.

New Zealand Ministry of Forestry. 1996. *Forestry sector issues: a post election briefing for the Ministry of Forestry.* Wellington.

New Zealand Ministry of Forestry. 1997. *A national exotic forest description.* Wellington.

Nousiainen, I.K. & Vesisenaho, T.J. 1998. Potential utilization of woodfuels in Finland. *In* P. Hakkila, M. Heino & E. Puranen, eds. *Wood fuels from conventional forestry.* Proceedings of the Workshop on Woodfuels from Conventional Forestry, Jasper, Alberta, Canada, 18 October 1997. Metsäntutkimuslaitoksen tiedonantoja 680 (Research Papers No. 680). Helsinki, Finland, Finnish Forest Research Institute.

Overseas Development Agency (ODA) (now Department for International Development – DFID). 1996. *Sharing forest management: key factors, best practice and ways forward.* Findings from ODA's review of participatory forest management. London, UK.

Poschen, P. 1997. Forests and employment – much more than meets the eye. In *Proceedings of the XI World Forestry Congress,* Vol. 4, p. 61-79. Rome, FAO.

Preston, L., ed. 1997. *Investing in mountains: innovative mechanisms and promising examples for financing conservation and sustainable development.* Franklin, West Virginia, USA, Mountain Forum/The Mountain Institute/FAO.

Price, M.F. 1998. Mountains: globally important ecosystems. *Unasylva,* 49(195): 3-12.

Rametsteiner, E., Schwarzbauer, P., Juslin, H., Kaernae, J., Cooper, R., Samuel, J., Becker, M. & Kuehn, T. 1998. *Proceedings of the Conference on Potential Markets for Certified Forest Products in Europe*, Brussels, Belgium, 13 March 1998. Joensuu, Finland, European Forestry Institute.

Reuters News Service. 1998. At least two dead in Russian forest fires. 21 September.

Reuters News Service. 1998. New steps to protect New York City drinking water. 8 August.

Schemo, D.J. 1998. Fires posing greater risk as Amazon rain forest grows drier. *New York Times*, 13 September.

Schimel, D., Alves, D., Enting, I., Heimann, M., Joos, F., Raynaud, D. & Wigley, T. 1996. Radiative forcing of climate change. *In* J.T. Houghton, L.G. Meira Filho, B.A. Callander, N. Harris, A. Kattenberg & K. Maskell, eds. *Climate change 1995 – the science of climate change.* Contribution of Working Group I to the Second Assessment Report of the Intergovernmental Panel on Climate Change. Cambridge, UK, Cambridge University Press.

Schwartzman, S. 1998. Burning up again in Brazilian Amazon. *INFOTERRA*, 16 September.

Schweithelm, J. 1998. *Public policies affecting forest fires in Asia-Pacific.* Paper presented to the Meeting on Public Policies Affecting Forest Fires, Rome, 28-30 October 1998.

Sist, P., Dykstra, D. & Fimbel, R. 1998. *Reduced-impact logging guidelines for lowland dipterocarp forests in Indonesia.* CIFOR Occasional Paper No. 15. Bogor, Indonesia, CIFOR.

Sizer, N. 1996. *Profit without plunder: reaping revenue from Guyana's tropical forests without destroying them.* Washington, DC, USA, World Resources Institute.

SkogForsk (Forestry Research Institute of Sweden). 1998. *CTI på virkesfordon ger bättre framkomlighet och större dragkraft.* Resultat No. 2. Uppsala, Sweden.

Third Ministerial Conference on the Protection of Forests in Europe/Ministry of Agriculture, Rural Development and Fisheries of Portugal. 1998a. Interim data of TBFRA 2000 as presented to the Third Ministerial Conference on Protection of Forests in Europe. In *Follow-up reports on the Ministerial Conferences on the Protection of Forests in Europe*, Vol. 2. Lisbon, Portugal.

Third Ministerial Conference on the Protection of Forests in Europe/Ministry of Agriculture, Rural Development and Fisheries of Portugal. 1998b. Status of sustainable forest management in Europe. In *Follow-up reports on the Ministerial Conferences on the Protection of Forests in Europe.* Lisbon, Portugal.

Thornqvist, T. 1998. *Handling of woodfuel at Södra Skogsenergi in Sweden.* Paper presented at IEA, Task 25, Workshop, Nokia, Finland, September 1998.

United Nations (UN). 1996. *World population prospects – the 1996 revision.* New York, NY, USA.

UN. 1997a. *World urbanization prospects – the 1996 revision.* New York, NY, USA.

UN. 1997b. *UN-ECE/FAO Temperate and Boreal Forest Resources Assessment 2000: terms and definitions.* New York, NY, USA and Geneva, Switzerland.

United Nations Environment Programme (UNEP)/World Conservation Monitoring Centre (WCMC). 1996. *Framework for information management in the context of the Convention on Biological Diversity.* Cambridge, UK, WCMC.

United States Department of Agriculture (USDA) Forest Service. 1996. *Activities related to poplar and willow cultivation, exploitation and utilization*, by J.G. Isebrands. Report of the National Poplar Commission of the USA to the 20th session of the International Poplar Commission, Budapest, Hungary, 1-4 October 1996.

USDA Forest Service. 1998. *Ice storm 1998. Interim status report.* Radnor, Pennsylvania, USA, Northeastern Area State and Private Forestry.

World Bank. 1997. *World development indicators 1997.* Washington, DC, USA.

World Conservation Monitoring Centre (WCMC). 1994. *Biodiversity data sourcebook*, ed. B. Groombridge. Cambridge, UK, World Conservation Press.

WCMC. 1996a. *The WWF world forest map.* Cambridge, UK.

WCMC. 1996b. *1996 IUCN red list of threatened animals*, ed. J. Baillie & B. Groombridge. Gland, Switzerland and Cambridge, UK, IUCN / World Conservation Union.

WCMC. 1998a. *The world list of threatened trees*, ed. S. Oldfield, C. Lusty, & A. MacKinven. Cambridge, UK, World Conservation Press.

WCMC. 1998b. *1997 IUCN red list of threatened plants*, ed. K.S. Walter & H.J. Gillett. Gland, Switzerland and Cambridge, UK, IUCN / World Conservation Union.

World Resources Institute (WRI)/UNEP/United Nations Development Programme (UNDP)/World Bank. 1996. *World Resources 1996-1997.* New York, NY, USA, Oxford University Press.

World Trade Organization (WTO). 1998. Report on the CTE meeting of 19-20 March 1998. *Trade and Environmental Bulletin*, 14 May.

World Wide Fund for Nature (WWF)/World Conservation Union (IUCN). 1994. *Centres of plant diversity. A guide and strategy for their conservation.* 3 vols. Cambridge, UK, IUCN. ◆

WHERE TO PURCHASE FAO PUBLICATIONS LOCALLY
POINTS DE VENTE DES PUBLICATIONS DE LA FAO
PUNTOS DE VENTA DE PUBLICACIONES DE LA FAO

• ***ANGOLA***
Empresa Nacional do Disco e de Publicações, ENDIPU-U.E.E.
Rua Cirilo da Conceição Silva, Nº 7
C.P. Nº 1314-C, Luanda

• ***ARGENTINA***
Librería Agropecuaria
Pasteur 743, 1028 Buenos Aires
Oficina del Libro Internacional
Av. Córdoba 1877, 1120 Buenos Aires
Correo electrónico: olilibro@satlink.com

• ***AUSTRALIA***
Hunter Publications
PO Box 404, Abbotsford, Vic. 3067
Tel.: 61 3 9417 5361
Fax: 61 3 9419 7154
E-mail: jpdavies@ozemail.com.au

• ***AUSTRIA***
Gerold Buch & Co.
Weihburggasse 26, 1010 Vienna

• ***BANGLADESH***
Association of Development Agencies in Bangladesh
House No. 1/3, Block F
Lalmatia, Dhaka 1207

• ***BELGIQUE***
M.J. De Lannoy
202, avenue du Roi, B-1060 Bruxelles
CCP 000-0808993-13
Mél.: jean.de.lannoy@infoboard.be

• ***BOLIVIA***
Los Amigos del Libro
Av. Heroínas 311, Casilla 450
Cochabamba;
Mercado 1315, La Paz

• ***BOTSWANA***
Botsalo Books (Pty) Ltd
PO Box 1532, Gaborone

• ***BRAZIL***
Fundação Getúlio Vargas
Praia do Botafogo 190, C.P. 9052
Rio de Janeiro
E-mail: valeria@sede.fgvrj.br
Núcleo Editora da Universidade Federal Fluminense
Rua Miguel de Frias 9
Icaraí-Niterói 24
220-000 Rio de Janeiro
Fundação da Universidade Federal do Paraná - FUNPAR
Rua Alfredo Bufrem 140, 30º andar
80020-240 Curitiba

• ***CAMEROUN***
CADDES
Centre Africain de Diffusion et Développement Social
B.P. 7317 Douala Bassa
Tél.: +237 43 37 83
Télécopie: +237 42 77 03

• ***CANADA***
Renouf Publishing
5369 chemin Canotek Road, Unit 1
Ottawa, Ontario K1J 9J3
Tel.: +1 613 745 2665
Fax: +1 613 745 7660
Website: www.renoufbooks.com
E-mail: renouf@fox.nstn.ca

• ***CHILE***
Librería - Oficina Regional, FAO
c/o FAO, Oficina Regional para América Latina y el Caribe (RLC)
Avda. Dag Hammarskjold, 3241
Vitacura, Santiago
Tel.: +56 2 33 72 314
Correo electrónico:
german.rojas@field.fao.org
Universitaria Textolibros Ltda.
Avda. L. Bernardo O'Higgins 1050
Santiago

• ***CHINA***
China National Publications Import & Export Corporation
16 Gongti East Road, Beijing 100020
Tel.: +86 10 6506 3070
Fax: +86 10 6506 3101
E-mail: cnpiec@public.3.bta.net.cn

• ***COLOMBIA***
INFOENLACE LTDA
Apartado Aéreo 34270
Santafé de Bogotá
Tel.: +57 1 2851779
Fax: +51 1 288 9882/255 2967
Correo electrónico:
infoenlace@gaitana.interred.net.co

• ***CONGO***
Office national des librairies populaires
B.P. 577, Brazzaville

• ***COSTA RICA***
Librería Lehmann S.A.
Av. Central, Apartado 10011
1000 San José
CINDE
Coalición Costarricense de Iniciativas de Desarrollo
Apartado 7170, 1000 San José
Correo electrónico:
rtacinde@sol.rassa.co.cr

• ***CÔTE D'IVOIRE***
CEDA
04 B.P. 541, Abidjan 04
Tél.: +225 22 20 55
Télécopie: +225 21 72 62

• ***CUBA***
Ediciones Cubanas
Empresa de Comercio Exterior de Publicaciones
Obispo 461, Apartado 605, La Habana

• ***CZECH REPUBLIC***
Artia Pegas Press Ltd
Import of Periodicals
Palác Metro, PO Box 825
Národní 25, 111 21 Praha 1

• ***DENMARK***
Munksgaard, Direct
Ostergate 26 A - Postbox 173
DK - 1005 Copenhagen K.
Tel.: +45 331 28570
Fax: +45 331 29387
URL address: www.munksgaardirect.dk
e-mail: direct@munksgaarddirect.dk

• ***REPÚBLICA DOMINICANA***
CUESTA - Centro del libro
Av. 27 de Febrero, esq. A. Lincoln
Centro Comercial Nacional
Apartado 1241, Santo Domingo
CEDAF - Centro para el Desarrollo Agropecuario y Forestal, Inc.
Calle José Amado Soler, 50 - Urban. Paraíso
Apartado Postal, 567-2, Santo Domingo
Correo electrónico: fda@Codetel.net.do
Sitio Web: http//www.fda.org.do
Tel.: +001 809 544-0616/544-0634/565-5603
Fax: +001 809 544-4727/567-6989

• ***ECUADOR***
Libri Mundi, Librería Internacional
Juan León Mera 851
Apartado Postal 3029, Quito
Correo electrónico:
librimul@librimundi.com.ec
Universidad Agraria del Ecuador
Centro de Información Agraria
Av. 23 de julio, Apartado 09-01-1248
Guayaquil
Librería Española
Murgeón 364 y Ulloa, Quito

• ***EGYPT***
The Middle East Observer
41 Sherif Street, Cairo
Tel.: +20 2 393 972
Fax: +20 2 360 6804
E-mail: book-order@meobserver.com.eg
Website: www.meobserver.com.eg

• ***ESPAÑA***
Librería Agrícola
Fernando VI 2, 28004 Madrid
Librería de la Generalitat de Catalunya
Rambla dels Estudis 118 (Palau Moja)
08002 Barcelona
Tel.: +34 93 302 6462
Fax: +34 93 302 1299
Mundi Prensa Libros S.A.
Castelló 37, 28001 Madrid
Tel.: +34 914 36 37 00
Fax: +34 915 75 39 98
Sitio Web: www.mundiprensa.com
Correo electrónico:
libreria@mundiprensa.es
Mundi Prensa - Barcelona
Consejo de Ciento 391,
08009 Barcelona
Tel.: +34 934 88 34 92
Fax: +34 934 87 76 59

• ***FINLAND***
Akateeminen Kirjakauppa
Subscription Services
PO Box 23, FIN-00371 Helsinki
Tel.: +358 9 121 4416
Fax: +358 9 121 4450

• ***FRANCE***
Editions A. Pedone
13, rue Soufflot, 75005 Paris
Lavoisier Tec & Doc
14, rue de Provigny
94236 Cachan Cedex
Site Web: www.lavoisier.fr
Mél.: livres@lavoisier.fr
Librairie du commerce international
10, avenue d'Iéna
75783 Paris Cedex 16
Site Web: www.cfce.fr
Mél.: pl@net-export.fr
WORLD DATA
10, rue Nicolas Flamand
75004 Paris
Tél.: +33 1 4278 0578
Télécopie: +33 1 4278 1472

• ***GERMANY***
Alexander Horn Internationale Buchhandlung
Friedrichstrasse 34
D-65185 Wiesbaden
Tel.: +49 6121 37 42 12
S. Toeche-Mittler GmbH
Versandbuchhandlung
Hindenburgstrasse 33
D-64295 Darmstadt
Tel.: +49 6151 33665
Fax: +49 6151 314 043
Website: www.booksell.com/triops
E-mail: triops@booksell.com
Uno Verlag
Poppelsdorfer Allee 55
D-53115 Bonn 1
Tel.: +49 228 94 90 20
Fax: +49 228 21 74 92
Website: www.uno-verlag.de
E-mail: unoverlag@aol.com

• ***GHANA***
SEDCO Publishing Ltd
Sedco House, Tabon Street
Off Ring Road Central, North Ridge
PO Box 2051, Accra
Readwide Bookshop Ltd
PO Box 0600 Osu, Accra
Tel.: +233 21 22 1387
Fax: +233 21 66 3347
E-mail: readwide@africaonline.cpm.gh

• ***GREECE***
Papasotiriou S.A.
35 Stournara Str., 10682 Athens
Tel.: +30 1 3302 980
Fax: +30 1 3648254

• ***GUYANA***
Guyana National Trading Corporation Ltd
45-47 Water Street, PO Box 308
Georgetown

• ***HAÏTI***
Librairie «A la Caravelle»
26, rue Bonne Foi
B.P. 111, Port-au-Prince

• ***HONDURAS***
Escuela Agrícola Panamericana
Librería RTAC
El Zamorano, Apartado 93, Tegucigalpa
Oficina de la Escuela Agrícola Panamericana en Tegucigalpa
Blvd. Morazán, Apts. Glapson
Apartado 93, Tegucigalpa

• ***HUNGARY***
Librotrade Kft.
PO Box 126, H-1656 Budapest
Tel.: +36 1 256 1672
Fax: +36 1 256 8727

• ***INDIA***
Allied Publisher Ltd
751 Mount Road
Chennai 600 002
Tel.: +91 44 8523938/8523984
Fax: +91 44 8520649
E-mail:
allied.mds@smb.sprintrpg.ems.vsnl.net.in
EWP Affiliated East-West Press PVT, Ltd
G-I/16, Ansari Road, Darya Gany
New Delhi 110002
Tel.: +91 11 3264 180
Fax: +91 11 3260 358
Oxford Book and Stationery Co.
Scindia House
New Delhi 110001
Tel.: +91 11 331 5310
Fax: +91 11 371 3275
Oxford Subscription Agency
Institute for Development Education
1 Anasuya Ave., Kilpauk
Madras 600 010
Periodical Expert Book Agency
G-56, 2nd Floor, Laxmi Nagar
Vikas Marg, Delhi 110092
Tel: +91 11 2215045/2150534
Fax: +91 11 2418599
Bookwell
Head Office:
2/72, Nirankari Colony, New Delhi - 110009
Tel.: +91 11 725 1283
Fax: +91 11 328 13 15
Sales Office:
24/4800, Ansari Road
Darya Ganj, New Delhi - 110002
Tel.: +91 11 326 8786
E-mail: bkwell@nde.vsnl.net.in

• ***IRAN***
The FAO Bureau, International and Regional Specialized Organizations Affairs
Ministry of Agriculture of the Islamic Republic of Iran
Keshavarz Bld, M.O.A., 17th floor
Teheran

• ***IRELAND***
Office of Public Work
4-5 Harcourt Road, Dublin 2

• ***ISRAEL***
R.O.Y. International
PO Box 13056, Tel Aviv 61130
E-mail: royil@netvision.net.il

WHERE TO PURCHASE FAO PUBLICATIONS LOCALLY
POINTS DE VENTE DES PUBLICATIONS DE LA FAO
PUNTOS DE VENTA DE PUBLICACIONES DE LA FAO

12/98

• ITALY
FAO Bookshop
Viale delle Terme di Caracalla
00100 Roma
Tel.: +39 06 5705 2313
Fax: +39 06 5705 3360
E-mail: publications-sales@fao.org
Libreria Commissionaria Sansoni S.p.A. - Licosa
Via Duca di Calabria 1/1
50125 Firenze
Tel.: +39 55 64 8 31
Fax: +39 55 64 12 57
E-mail: licosa@ftbcc.it
Libreria Scientifica Dott. Lucio de Biasio "Aeiou"
Via Coronelli 6, 20146 Milano

• JAPAN
Far Eastern Booksellers (Kyokuto Shoten Ltd)
12 Kanda-Jimbocho 2 chome
Chiyoda-ku - P.O. Box 72
Tokyo 101-91
Tel.: +81 3 3275 8585
Fax: +81 3 3275 0656
Maruzen Company Ltd
PO Box 5050
Tokyo International 100-31
Tel.: +81 33 278 1894
Fax: +81 3 3 278 1895
E-mail: h_sugiyama@maruzen.co.jp

• KENYA
Text Book Centre Ltd
Kijabe Street
PO Box 47540, Nairobi
Tel.: +254 2 330 342
Fax: +254 2 22 57 79
Inter Africa Book Distribution
Kencom House, Moi Avenue
PO Box 73580, Nairobi
Tel.: +254 2 211 184
Fax: +254 2 22 3 5 70
Legacy Books
Mezzanine 1, Loita House, Loita Street
Nairobi, PO Box 68077
Tel.: +254 2 303853
Fax: +254 2 330854

• LUXEMBOURG
M.J. De Lannoy
202, avenue du Roi
B-1060 Bruxelles (Belgique)
Mél.: jean.de.lannoy@infoboard.be

• MADAGASCAR
Centre d'Information et de Documentation Scientifique et Technique
Ministère de la recherche appliquée au développement
B.P. 6224 Tsimbazaza, Antananarivo

• MALAYSIA
Electronic products only:
Southbound Sendirian Berhad Publishers
9 College Square, 01250 Penang

• MALI
Librairie Traore
Rue Soundiata Keita X 115
B.P. 3243, Bamako

• MAROC
La Librairie Internationale
70 Rue T'ssoule
PO Box 302 (RP), Rabat
Tél./Télécopie: 212 7 75 01 83

• MÉXICO
Librería, Universidad Autónoma de Chapingo
56230 Chapingo
Libros y Editoriales S.A.
Av. Progreso Nº 202-1º Piso A
Apartado. Postal 18922
Col. Escandón, 11800 México D.F.
Mundi Prensa Mexico, S.A.
Río Pánuco, 141 Col. Cuauhtémoc
C.P. 06500, México, DF
Tel.: +52 5 533 56 58
Fax: +52 5 514 67 99
Correo electrónico:
1015452361@compuserve.com

• NETHERLANDS
Roodveldt Import b.v.
Brouwersgracht 288
1013 HG Amsterdam
E-mail: roodboek@euronet.nl
Tel.: +31 20 622 80 35
Fax: +31 20 625 54 93
Swets & Zeitlinger b.v.
PO Box 830, 2160 Lisse
Heereweg 347 B, 2161 CA Lisse
E-mail: infono@swets.nl
Website: www.swets.nl

• NEW ZEALAND
Legislation Services
PO Box 12418
Thorndon, Wellington
E-mail: gppmjxf@gp.co.nz
Oasis Official
PO Box 3627, Wellington
Tel.: +64 4 499 1551
Fax: +64 4 499 1972
E-mail: oasis@clear.net.nz
Website: www.oasisbooks.co.nzl

• NICARAGUA
Librería HISPAMER
Costado Este Univ. Centroamericana
Apartado Postal A-221, Managua

• NIGERIA
University Bookshop (Nigeria) Ltd
University of Ibadan, Ibadan

• PAKISTAN
Mirza Book Agency
65 Shahrah-e-Quaid-e-Azam
PO Box 729, Lahore 3

• PARAGUAY
Librería Intercontinental Editora e Impresora S.R.L.
Caballero 270 c/Mcal Estigarribia
Asunción

• PERÚ
INDEAR
Jirón Apurímac 375, Casilla 4937
Lima 1
Universidad Nacional «Pedro Ruiz Gallo»
Facultad de Agronomía, A.P. 795
Lambayeque (Chiclayo)

• PHILIPPINES
International Booksource Center, Inc.
Room 720, Cityland 10, Tower 2
H.V. de la Costa, Cor. Valero St
Makati, Metro Manila
Tel.: +63 2 817 9676
Fax: +63 2 817 1741

• POLAND
Ars Polona
Krakowskie Przedmiescie 7
00-950 Warsaw

• PORTUGAL
Livraria Portugal, Dias e Andrade Ltda.
Rua do Carmo, 70-74
Apartado 2681, 1200 Lisboa Codex

• SINGAPORE
Select Books Pte Ltd
03-15 Tanglin Shopping Centre
19 Tanglin Road, Singapore 1024
Tel.: +65 732 1515
Fax: +65 736 0855

• SLOVAK REPUBLIC
Institute of Scientific and Technical Information for Agriculture
Samova 9, 950 10 Nitra
Tel.: +421 87 522 185
Fax: +421 87 525 275
E-mail: uvtip@nr.sanet.sk

• SOMALIA
Samater
PO Box 936, Mogadishu

• SOUTH AFRICA
David Philip Publishers (Pty) Ltd
PO Box 23408, Claremont 7735
Tel.: Cape Town +27 21 64 4136
Fax: Cape Town+ 27 21 64 3358
E-mail: dpp@iafrica.com
Website: www.twisted.co.za

• SRI LANKA
M.D. Gunasena & Co. Ltd
217 Olcott Mawatha, PO Box 246
Colombo 11

• SUISSE
Buchhandlung und Antiquariat Heinimann & Co.
Kirchgasse 17, 8001 Zurich
UN Bookshop
Palais des Nations
CH-1211 Genève 1
Site Web: www.un.org
Van Diermen Editions Techniques ADECO
41 Lacuez, CH-1807 Blonzy

• SURINAME
Vaco n.v. in Suriname
Domineestraat 26, PO Box 1841
Paramaribo

• SWEDEN
Wennergren Williams AB
PO Box 1305, S-171 25 Solna
Tel.: +46 8 705 9750
Fax: +46 8 27 00 71
E-mail: mail@wwi.se
Bokdistributören
c/o Longus Books Import
PO Box 610, S-151 27 Södertälje
Tel.: +46 8 55 09 49 70
Fax: +46 55 01 76 10
E-mail: lis.ledin@hk.akademibokhandeln.se

• THAILAND
Suksapan Panit
Mansion 9, Rajdamnern Avenue
Bangkok

• TOGO
Librairie du Bon Pasteur
B.P. 1164, Lomé

• TUNISIE
Société tunisienne de diffusion
5, avenue de Carthage, Tunis

• TURKEY
DUNYA INFOTEL
100. Yil Mahallesi
34440 Bagcilar, Istanbul
Tel.: +90 212 629 0808
Fax: +90 212 629 4689
E-mail: dunya@dunya-gazete.com.tr
Website: http://www.dunya.com

• UGANDA
Fountain Publishers Ltd
PO Box 488, Kampala
Tel.: +256 41 259 163
Fax: +256 41 251 160

• UNITED KINGDOM
The Stationery Office
51 Nine Elms Lane
London SW8 5DR
Tel.: +44 171 873 9090 (orders)
+44 171 873 0011 (inquiries)
Fax: +44 171 873 8463
and through The Stationery Office Bookshops
E-mail: postmaster@theso.co.uk
Website: www.the-stationery-office.co.uk
Electronic products only:
Microinfo Ltd
PO Box 3, Omega Road
Alton, Hampshire GU34 2PG
Tel.: +44 1420 86 848
Fax: +44 1420 89 889
Website: www.microinfo.co.uk
E-mail: emedia@microinfo.co.uk

• UNITED STATES
Publications:
BERNAN Associates (ex UNIPUB)
4611/F Assembly Drive
Lanham, MD 20706-4391
Toll-free: +1 800 86534
Fax: +1 800 865 3450
Website: www.bernan.com
E-mail: query@bernan.com
UN Bookshop
The United Nations Bookshop
General Assembly Building Room 32
New York, N.Y. 10017
Tel.: +1 212 963 7680
Fax: +1 212 963 4910
Website: www.un.org
E-mail: bookshop@un.org
Periodicals:
Ebsco Subscription Services
PO Box 1943
Birmingham, AL 35201-1943
Tel.: +1 205 991 6600
Fax: +1 205 991 1449
The Faxon Company Inc.
15 Southwest Park
Westwood, MA 02090
Tel.: 6117-329-3350
Telex: 95-1980
Cable: FW Faxon Wood

• URUGUAY
Librería Agropecuaria S.R.L.
Buenos Aires 335, Casilla 1755
Montevideo C.P. 11000

• VENEZUELA
Fundación La Era Agrícola
Calle 31 Junín Qta Coromoto 5-49
Apartado 456, Mérida
Fudeco, Librería
Avenida Libertador-Este
Ed. Fudeco, Apartado 254
Barquisimeto C.P. 3002, Ed. Lara
Tel.: +58 51 538 022
Fax: +58 51 544 394
Librería FAGRO
Universidad Central de Venezuela (UCV)
Maracay
Librería Universitaria, C.A.
Av. 3, entre Calles 29 y 30
Nº 29-25 Edif. EVA, Mérida
Fax: +58 74 52 0956
Tamanaco Libros Técnicos S.R.L.
Centro Comercial Ciudad Tamanaco
Nivel C-2, Caracas
Tel.: +58 2 261 3344/261 3335
Tecni-Ciencia Libros S.A.
Torre Phelps-Mezzanina
Plaza Venezuela
Apartado Postal: 20.315, 1020 Caracas
Tel.: +58 2 782 8698/781 9945
E-mail: tchlibros@ibm.net

• ZIMBABWE
Grassroots Books
The Book Café
Fife Avenue, Harare;
61a Fort Street, Bulawayo
Tel.: +263 4 79 31 82
Fax: +263 4 72 62 43

• Other countries/Autres pays/ Otros países
Sales and Marketing Group
Information Division, FAO
Viale delle Terme di Caracalla
00100 Rome, Italy
Tel.: +39 06 57051
Fax: +39 06 5705 3360
E-mail: publications-sales@fao.org